Man Kam Kwong   Anton Zettl

Norm Inequalities for Derivatives and Differences

Springer-Verlag
Berlin Heidelberg New York
London Paris Tokyo
Hong Kong Barcelona
Budapest

Authors

Man Kam Kwong
Mathematics and Computer Science Division
Argonne National Laboratory
Argonne, IL 60439
USA

Anton Zettl
Department of Mathematical Sciences
Northern Illinois University
DeKalb, IL 60115
USA

Mathematics Subject Classification (1991): 2602, 26D15, 3902, 39A12, 39A70, 39B72, 47A30, 47B39

ISBN 3-540-56387-3 Springer-Verlag Berlin Heidelberg New York
ISBN 0-387-56387-3 Springer-Verlag New York Berlin Heidelberg

© Springer-Verlag Berlin Heidelberg 1992
Printed in Germany

Typesetting: Camera ready by author
Printing and binding: Druckhaus Beltz, Hemsbach/Bergstr.
46/3140-543210 - Printed on acid-free paper

Contents

Preface

Edmund Landau's 1913 paper "Einige Ungleichungen für zweimal differenzierbare Funktionen", based on earlier work of Hardy and Littlewood, initiated a vast and fruitful research activity involving the study of the relationship between the norms of (i) a function and its derivatives and (ii) a sequence and its differences. These notes are an attempt to give a connected account of this effort. Detailed elementary proofs of basic inequalities are given. These are accessible to anyone with a background of advanced calculus and a rudimentary knowledge of the L^p and l^p spaces, yet the reader will be brought to the frontier of knowledge regarding several aspects of these problems. Many open questions are raised.

We thank Judy Beumer and Diane Keding for their careful typing of this difficult manuscript. Special thanks are also due Mari-Anne Hartig for never losing patience with our repeated requests for changes and without whose expertise in constructing graphs, tables, getting cross-references right, etc. this final version would not have been possible.

Introduction

The norm of a function y may not be related to the norm of its derivative y'. One may be large while the other is small. More precisely, given any positive numbers u_0 and u_1, there exists a differentiable function y satisfying

$$\|y\| = u_0 \text{ and } \|y'\| = u_1. \tag{0.1}$$

This is true and easy to prove in particular for the classical p-norms:

$$\|y\|_p^p = \int_J | y(t) |^p \, dt, \ 1 \le p < \infty$$
$$\|y\|_\infty = \text{ess. sup } | y(t) |, \ p = \infty$$

where J is any bounded or unbounded interval of the real line.

Given another positive number u_2: Does there exist a function y which satisfies, in addition to (0.1), also

$$\|y\| = u_2 \ ?$$

The answer is *no*, thus giving rise to another question: How are the norms of a function y and its derivatives related to each other? It is this question we study in this monograph.

Of primary interest is the classical inequality

$$\|y^{(k)}\|^n \ \le \ K \, \|y\|^{n-k} \, \|y^{(n)}\|^k$$

often associated with the names of Landau, Hardy and Littlewood, Kolmogorov, among others, and its discrete analogue

$$\|\triangle^k x\|^n \ \le \ C \, \|x\|^{n-k} \, \|\triangle^n x\|^k.$$

Our goal is to give a basically self-contained exposition requiring only a background of advanced calculus and the basics of Lebesque integration theory, yet we aim to bring the reader to the frontier of knowledge for some aspects of these inequalities. Many results obtained here are from papers less than 15 years old; in the discrete case, less than 10 years. Some are more recent than that. A lot of open problems are mentioned, many of which, we believe, are "accessible". An extensive bibliography is also provided which includes some of the vast Soviet literature on this topic.

Chapter 1

Unit Weight Functions

In this chapter, we discuss the values that the norm of a function and its derivatives can assume. Considered are the classical L^p norms with unit weight. Some basic inequalities are discussed, including those often associated with the names of Landau, Hardy-Littlewood and Kolmogorov.

Although the subject matter treated in this chapter is very classical and the methods used are elementary, there are some results here which do not seem to have been published before.

1.1 The Norms of y and $y^{(n)}$

The classic p-norms are defined by

$$\|y\|_p = \left(\int_J |y(t)|^p \, dt \right)^{1/p}, \quad 1 \le p < \infty,$$

$$\|y\|_\infty = \text{ess sup } |y(t)|, \quad t \in J, \quad p = \infty.$$

Here J is any nondegenerate interval of the real line, bounded or unbounded.

The set of equivalence classes (with respect to Lebesgue measure) of functions whose p norms are finite is the classical Banach space $L^p(J)$, $1 \le p \le \infty$. Below, $y^{(n)}$ denotes the nth derivative of y and $y^{(n)} \in L^q(J)$ means that $y^{(n-1)}$ is absolutely continuous on any compact subinterval of J, so that $y^{(n)}$ exists a.e. and is locally integrable, and $\|y^{(n)}\|_q$ is finite. The symbol $\|y\|_{p,J}$ will be used when we wish to emphasize the dependence of the norm on the interval J. If p or J is fixed in a result or argument we may merely use the symbol $\|y\|$. Throughout the book, p and q are assumed to satisfy $1 \le p, q \le \infty$.

We define $W_{p,q}^n(J)$ to be the subspace of $L^p(J)$ consisting of functions $y \in L^p(J)$ such that $y^{(n)} \in L^q(J)$. No integrability conditions are imposed on $y^{(k)}$ for $1 \le k < n$.

Two exponents p and q $(1 \le p, q \le \infty)$ are said to be conjugate to each other if $p^{-1} + q^{-1} = 1$. We follow the usual convention that $p = 1$ when $q = \infty$ and $p = \infty$ when $q = 1$.

The symbol $C^n(J)$ denotes the set of complex-valued functions with a continuous n^{th} derivative on J, $C^\infty(J)$ is the set of infinitely differentiable functions on J, and $C_0^\infty(J)$ is the set of infinitely differentiable functions with compact support in the interior of J.

In this monograph we are concerned with various inequalities among the norms of derivatives of a function. Our first result ensures the existence of a function whose norm and whose n^{th} derivative's norm are respectively equal to two arbitrary given positive numbers. This is not at all surprising and the proof is not difficult.

Theorem 1.1 *Let $1 \le p, q \le \infty$, let n be a positive integer, and let J be any interval on the real line, bounded or unbounded. Given any positive numbers u and v there exists a function $y \in C^n(J)$ such that*

$$\|y\|_p = u, \quad \|y^{(n)}\|_q = v. \tag{1.1}$$

Proof.

Case 1. $1 \le p, q < \infty$, $J = R = (-\infty, \infty)$.

Choose y in $C_0^\infty(R)$ such that $y \ge 0$ but not identically zero and y has compact support. Consider

$$y_{ab}(t) = ay(bt), \quad a > 0, \quad b > 0$$

and note that with $x = bt$

$$\|y_{ab}\|_p^p = \int_R a^p |y(bt)|^p dt = a^p b^{-1} \int_R |y(x)|^p dx = a^p b^{-1} \|y\|_p^p$$

$$\|y_{ab}^{(n)}\|_q^q = \int_R a^q b^{nq} |y^{(n)}(bt)|^q dt = a^q b^{nq-1} \int_R |y^{(n)}(x)|^q dx$$

$$= a^q b^{nq-1} \|y^{(n)}\|_q^q.$$

Choose b such that

$$b^{nq-1+q/p} = v^q \|y\|_p^q u^{-q} \|y^{(n)}\|_q^{-q} \tag{1.2}$$

(observe that $y^{(n)}$ is not identically zero since y has compact support and is not the zero function), and then choose $a = b^{1/p} u / \|y\|_p$.

The proof for $J = R^+ = (0, \infty)$ is similar. Since translation and reflection preserve norms, all other half-line cases $(-\infty, a)$ or (a, ∞), $-\infty < a < \infty$, reduce to R^+.

Case 2. $1 \leq p, q < \infty$, $J = (a, b)$, $-\infty < a < b < \infty$.

First consider the case $J = (0, 1)$. Define

$$Q(y) = \|y^{(n)}\|_q / \|y\|_p, \quad y \in W^n_{p,q}(J) = X.$$

Since the norm is a continuous function from X into $[0, \infty)$ it follows that Q is a continuous function from $X - \{0\}$ into $[0, \infty)$. We show that Q is onto. Let $Q(y) = \alpha$, $Q(z) = \beta$, $0 \leq \alpha < \beta$. Then y and z are linearly independent. Thus $S = \text{span}\{y, z\} - \{0\}$ is a two-dimensional connected subset of X. Since the continuous image of a connected set is connected, it follows that $[\alpha, \beta] \subset \text{range } Q$. By letting $y(t) = 1$ we see that we can choose $\alpha = 0$. To show that β can be chosen arbitrarily large, consider

$$y(t) = t^d, \quad 0 < t < 1.$$

Then

$$\|y\|_p^p = \int_0^1 t^{pd} dt = 1/(pd + 1)$$

$$\|y^{(n)}\|_q^q = h(d) \int_0^1 t^{(d-n)q} dt = h(d)/(q(d-n) + 1),$$

where

$$h(d) = d(d-1) \ldots [d - (n-1)].$$

Thus

$$Q(y) = (h(d))^{1/q} (pd + 1)^{1/p} (q(d-n) + 1)^{-1/q} \to \infty \text{ as } d \to \infty.$$

We conclude that the range of Q is $[0, \infty)$. Let $r = v/u$. ¿From the above argument we know there is a $y \in W^n_{p,q}(0, 1)$ such that $Q(y) = r$. Choose the constant c so that $\|cy\|_p = u$, then $\|cy^{(n)}\|_q = v$. This completes the proof for the case $J = (0, 1)$. The general case of a bounded interval J follows from this case and a transformation of the form $t \to ct + d = x$.

The proofs of the remaining cases ($p = \infty$ or $q = \infty$, J bounded or unbounded) are left to the reader as exercises. $\square$

The proof of Theorem 1.1 is presented as above to point out the difference between the cases of unbounded and bounded J. In the first case, the technique of "horizontal scaling" (change of the independent variable) is a very useful tool, which is not available in the second case.

1.2   The norms of y, $y^{(k)}$, and $y^{(n)}$

The special case $p = q$, $n = 1$ of Theorem 1.1 says that the norm of a function and its derivative on any interval, bounded or unbounded, can assume arbitrary positive values. Can the norms of y, y', and y'' assume arbitrary positive values? Below we will see that the answer is no. But first we discuss some preliminary results.

Lemma 1.1 *Let $1 \leq p \leq \infty$. Assume $J = [a, b]$ is a compact interval of length $L = b - a$. If $y \in L^p(J)$ and $y^{(k)}$ exists on J and*

$$\alpha_k = \alpha_k(J) = \inf |y^{(k)}(t)|, \quad t \in J; \tag{1.3}$$

then

$$\alpha_k(J) \leq A \, \|y\|_{p,J} \quad k = 1, 2, 3, \ldots, \tag{1.4}$$

where A is a constant independent of y given by

$$A = A(k, p, L) = 2^k \cdot 3^{x(k)} L^{-k-1/p}, \tag{1.5}$$

where $x(k)$ is defined recursively by

$$x(1) = 1/p, \quad x(k+1) = x(k) + k + 1/p, \quad k = 1, 2, 3, \ldots. \tag{1.6}$$

Proof. The proof uses a "triple interval" argument and induction on k.

Case 1. $p = \infty$.

Divide J into three equal subintervals: $J_1 = [a, a + L/3]$, $J_2 = [a + L/3, \quad a + 2L/3]$, and $J_3 = [a + 2L/3, b]$. By the mean value theorem, for any t_1 in J_1 and t_3 in J_3 there exists $t^* \in (t_1, t_3)$ such that

$$
\begin{aligned}
\alpha_1 &\leq \ |y'(t^*)| = |(y(t_1) - y(t_3))/(t_1 - t_3)| \\
&\leq \ 3L^{-1}(|y(t_1)| + |y(t_3)|) \leq 6L^{-1}\|y\|_{\infty,J}.
\end{aligned}
$$

This establishes the case $k = 1$. To establish the inductive step it is convenient to use the notation $\alpha_k(J)$ to denote the dependence of α_k on the interval J. For the sake of clarity we consider first the case $k = 2$. Using the above notation, choose t_i in J_i so that $\alpha_1(J_i) = |y'(t_i)|$, $i = 1, 3$. By the mean value theorem we have, for some $t^* \in (t_1, t_3)$

$$
\begin{aligned}
\alpha_2 &\leq \ |y''(t^*)| = |(y'(t_1) - y'(t_3))/(t_1 - t_3)| \\
&\leq \ 3L^{-1}(\alpha_1(J_1) + \alpha_1(J_3)) \\
&\leq \ 3L^{-1}(6(L/3)^{-1}\|y\|_{\infty,J_1} + 6(L/3)^{-1}\|y\|_{\infty,J_3}) \\
&\leq \ 3 \cdot 6^2 \cdot L^{-2}\|y\|_{\infty,J}.
\end{aligned}
$$

Assume (1.4), (1.5) hold in the kth stage of our induction process. Let $J = J_1 \cup J_2 \cup J_3$ as above in step $k = 1$. Then using the mean value theorem again with $t_1 \in J_1$, $t_3 \in J_3$ chosen so that $\alpha_k(J_i) = |y^{(k)}(t_i)|$, $i = 1,3$ we get

$$\begin{aligned}
\alpha_{k+1}(J) &\leq |y^{(k+1)}(t^*)| = |(y^{(k)}(t_1) - y^{(k)}(t_3))/(t_1 - t_3)| \\
&\leq L^{-1}3(\alpha_k(J_1) + \alpha_k(J_3)) \\
&\leq 3L^{-1}2^k 3^{x(k)}(L/3)^{-k}(\|y\|_{\infty,J_1} + \|y\|_{\infty,J_3}) \\
&\leq L^{-k-1}2^{k+1}3^{1+x(k)+k}\|y\|_{\infty,J}.
\end{aligned}$$

This completes the proof of (1.4), (1.5) for $p = \infty$.

Case 2. $1 \leq p < \infty$.

Let $p^{-1} + q^{-1} = 1$. Let $J = J_1 \cup J_2 \cup J_3$ be as in Case 1 above. Choose $t_i \in J_i$ such that $|y(t_i)| = \min |y(t)|$, $t \in J_i$, $i = 1,3$. ¿From the mean value theorem and Hölder's inequality we have for some t^* between t_1 and t_3

$$\alpha_1 \leq |y(t^*)| \leq L^{-1}3(|y(t_1)| + |y(t_3)|)$$

and

$$\begin{aligned}
L|y(t_i)|/3 &= \int_{J_1} |y(t_i)|dt \leq \int_{J_i} |y(t)|dt \\
&\leq (L/3)^{1/q}\|y\|_{p,J_i}, \quad i = 1,3.
\end{aligned}$$

¿From these inequalities

$$\alpha_1 \leq 3^{1-1/q}L^{1/q-2}2\|y\|_{p,J}.$$

This is (1.4), (1.5) for $k = 1$. Assume (1.4), (1.5) hold for k. Decompose J as above and choose $t_i \in J_i$ such that $|y^{(k)}(t_i)| = \inf |y^{(k)}(t)|$ for $t \in J_i$, $i = 1,3$. Then, as above,

$$\begin{aligned}
\alpha_{k+1}(J) &\leq |y^{(k+1)}(t^*)| \leq 3L^{-1}(|y^{(k)}(t_1)| + |y^{(k)}(t_3)|) \\
&= 3L^{-1}(\alpha_k(J_1) + \alpha_k(J_3)) \\
&\leq 3L^{-1}2(2^k 3^{x(k)}L^{1/q-k})\|y\|_{p,J},
\end{aligned}$$

and the proof of Lemma 1.1 is complete. □

Lemma 1.2 *Let* $1 \leq p,q,r \leq \infty$, $l(J) = L < \infty$. *If* $y \in L^p(J)$ *and* $y'' \in L^r(J)$ *then* $y' \in L^q(J)$ *and*

$$\|y'\|_q \leq AL^{1/r'+1/q}\|y''\|_r + BL^{-1-1/p+1/q}\|y\|_p, \tag{1.7}$$

where $1/r' + 1/r = 1$ *and*

$$A = 2^{1-1/q}, \quad B = 2^{2-1/q} \cdot 3^{1/p}. \tag{1.8}$$

Proof. We may assume, without loss of generality, that J is compact. With the notation of Lemma 1.1 we have

$$|y'(t)| \leq \alpha_1 + \left| \int_{t_1}^t y'' \right| \leq 2 \cdot 3^{1/p} L^{-1-1/p} \|y\|_p + L^{1/r'} \|y''\|_r. \tag{1.9}$$

If $q = \infty$, (1.7), (1.8) follow from (1.9) since $1 < 2$, $2 \cdot 3^{1/p} < B$. If $q < \infty$, we obtain from (1.9)

$$\begin{aligned} \int_J |y'(t)|^q dt &\leq 2^{q-1} (2^q \cdot 3^{q/p} L^{-q-q/p+1} \|y\|_p^q + L^{q/r'+1} \|y''\|_r^q) \\ &= A^q L^{q/r'+q} \|y''\|_r^q + B^q L^{-q-q/p+1} \|y\|_p^q \end{aligned} \tag{1.10}$$

using

$$a^s + b^s \leq (a+b)^s \leq 2^{s-1}(a^s + b^s), \quad 1 \leq s, \quad a > 0, \quad b > 0. \tag{1.11}$$

Now (1.7) follows from (1.10) and the second half of the elementary inequality

$$2^{s-1}(a^s + b^s) \leq (a+b)^s \leq a^s + b^s, \quad 0 \leq s \leq 1, \quad a > 0, \quad b > 0. \quad \square \tag{1.12}$$

Lemma 1.3 *Let $1 \leq p \leq \infty$, $l(J) = L \leq \infty$. Given $\epsilon > 0$ there exists a $K(\epsilon) > 0$ such that if $y \in L^p(J)$, y' is locally absolutely continuous on J, $y'' \in L^p(J)$ then $y' \in L^p(J)$ and*

$$\|y'\|_p \leq \epsilon \|y''\|_p + K(\epsilon) \|y\|_p. \tag{1.13}$$

Furthermore, for fixed ϵ, $K(\epsilon)$ can be chosen to be a nonincreasing function of the length of J.

Proof. We consider $p < \infty$ first.

Case 1. Assume $L < \infty$.

Let $\epsilon > 0$. If $L \leq \epsilon/A$ then (1.13) follows from (1.7) with

$$K(\epsilon) = BL^{-1}, \quad B = 2^{2-1/p}. \tag{1.14}$$

If $\epsilon_1 = \epsilon/A < L < \infty$ let $J = \cup_{i=1}^n J_i$, where J_i are nonoverlapping, $l(J_1) = \epsilon_1/2$, $i = 1, \ldots, n-1$, and $\epsilon_1/2 \leq l(J_n) \leq \epsilon_1$. Apply (1.10) to J_i, $i = 1, \ldots, n-1$ with L replaced by $\epsilon_1/2 = \epsilon/2A$, and q, r by p, we get

$$\int_I |y'|^p \leq \epsilon^p 2^{-p} \int_I |y''|^p + (2AB)^p \epsilon^{-p} \int_I |y|^p \leq \epsilon^p \int_I |y''|^p + (2AB)^p \epsilon^{-p} \int_I |y|^p \tag{1.15}$$

holding on each interval $I = J_i$, $i = 1, \ldots, n-1$. On $I = J_n$ we get from (1.10)

$$\begin{aligned} \int_I |y'|^p &\leq \epsilon^p \int_I |y''|^p + B^p L^{-p} \int_I |y|^p \\ &\leq \epsilon^p \int_I |y''|^p + (2AB)^p \epsilon^{-p} \int_I |y|^p. \end{aligned} \tag{1.16}$$

Summing inequalities (1.15) and (1.16) over all the intervals J_i, $i = 1, \ldots, n$ and then taking the pth root we obtain (1.13) with

$$K(\epsilon) = 2AB\epsilon^{-1}. \tag{1.17}$$

Case 2. $L = \infty$. Let $J = \cup_{i=1}^{\infty} J_i$, where J_i are nonoverlapping intervals each of length ϵ_1. Proceeding as above we get inequality (1.15) on each interval $I = J_i$, $i = 1, 2, 3, \ldots$. Summing over the intervals J_i and taking the pth root yields (1.13) with $K(\epsilon)$ given by (1.17).

For fixed ϵ it is clear that $K(\epsilon)$, chosen according to (1.14) and (1.17) is a nonincreasing function of the length of the interval J. This completes the proof for $p < \infty$. The modifications needed with $p = \infty$ are straightforward and hence omitted. $\square$

Theorem 1.2 *Let $1 \le p \le \infty$, let n, k be integers with $1 \le k < n$, and let J be any interval of the real line, bounded or unbounded. Given any $\epsilon > 0$ there exists a positive $K(\epsilon)$ such that if $y \in L^p(J)$, $y^{(n-1)}$ is locally absolutely continuous and $y^{(n)} \in L^p(J)$ (i.e. $y \in W_{p,p}^n(K)$) then $y^{(k)} \in L^p(J)$ and*

$$\|y^{(k)}\|_p \le \epsilon \|y^{(n)}\|_p + K(\epsilon) \|y\|_p. \tag{1.18}$$

Furthermore, for a given $\epsilon > 0$ the constant $K(\epsilon)$ can be chosen to be a non-increasing function of the length of the interval J.

Proof. The proof is by induction on n. Since p is fixed throughout the proof we will suppress the subscript p on the norm symbol. The case $n = 2$ is Lemma 1.3. Assume Theorem 1.2 is true for $n = N$ (and $k = 1, 2, \ldots, N-1$) and suppose that $y \in W_{p,p}^{N+1}(J)$. We need to show that $y^{(k)} \in L^p(J)$ for $1 \le k < N+1$ and (1.18) holds with $n = N+1, k \le N$. Note that it does not follow immediately from the induction hypothesis that $y^{(k)} \in L^p(J)$ for $k = 1, \ldots, N$. But $y^{(k)} \in L^p(I)$ for any compact subinterval I of J since $y^{(k)}$ is absolutely continuous on I, $k = 1, \ldots, N$. Hence by Lemma 1.3 given $\epsilon_1 > 0$ there exists a $K(\epsilon_1) > 0$ such that

$$\begin{aligned}
\|y^{(N)}\|_I &\le \epsilon_1 \|y^{(N+1)}\|_I + K(\epsilon_1)\|y^{(N+1)}\|_I \\
&\le \epsilon_1 \|y^{(N+1)}\|_J + K(\epsilon_1)(\epsilon_2 \|y^{(N)}\|_I + K(\epsilon_2)\|y\|_I).
\end{aligned}$$

Here we used the inductive hypothesis in the last step. Rearranging terms we get

$$(1 - K(\epsilon_1)\epsilon_2 \|y^{(N)}\|_I \le \epsilon_1 \|y^{(N+1)}\|_J + K(\epsilon_1)K(\epsilon_2)\|y\|_I.$$

Choose $\epsilon_1 < \epsilon/2$ and ϵ_2 such that $1/2 \le 1 - K(\epsilon_1)\epsilon_2 < 1$. Then dividing by $1 - K(\epsilon_1)\epsilon_2$ and using $\|y\|_I \le \|y\|_J$ we get

$$\|y^{(N)}\|_I \le \epsilon \|y^{(N+1)}\|_J + 2K(\epsilon_1)K(\epsilon_2)\|y\|_J. \tag{1.19}$$

Since this inequality holds for each compact subinterval I of J we conclude that $y^{(N)}$ is in $L^p(J)$ and (1.18) holds with $n = N + 1$ and $K(\epsilon) = 2K(\epsilon_1)K(\epsilon_2)$. The rest of the argument follows from (1.19), the induction hypothesis and similar computations.

The furthermore part of Theorem 1.2 follows from the choice $K(\epsilon) = 2K(\epsilon_1)K(\epsilon_2)$ and the fact that $K(\epsilon_1)$ and $K(\epsilon_2)$ both can be chosen as nonincreasing functions of the length of J; $K(\epsilon_1)$ by Lemma 1.3 and $K(\epsilon_2)$ by the induction hypothesis, and the proof of Theorem 1.1 is complete. $\square$

In Theorem 1.2 let

$$\mu_1 = \inf(\max(\epsilon, K(\epsilon)), \quad 0 < \epsilon < \infty \tag{1.20}$$

then

$$\|y^{(k)}\|_p \le \mu_1 \left[\|y\|_p + \|y^{(n)}\|_p \right]. \tag{1.21}$$

Let $\mu = \mu(k, n, p, J)$ denote the best, i.e., smallest constant in (1.21). The exact value of μ is not known except in a few special cases. We will mention some of these in Section 3. Clearly $\mu \le \mu_1$. ¿From the proof of Theorem 1.2 one does not expect equality to hold here. In fact it may happen that $\mu < \mu_1$ even if the best value for $K(\epsilon)$ for all $\epsilon > 0$ is known in (1.18). For the case $k = 1$, $n = 2$, $p = 2$, $J = [a, b]$, Redheffer [1963] has found the best constant $K(\epsilon)$ for each $\epsilon > 0$:

$$K(\epsilon) = 1/\epsilon + 12/(b-a)^2. \tag{1.22}$$

This yields $\mu_1 \approx 12.08$ for $b = 2, a = 1$. On the other hand, Phong [1981] developed an algorithm for the computation of μ according to which $\mu(1, 2, 3, [0, 1]) = 6.45 < 12$. (This algorithm of Phong's seems not to have been implemented in the case of a bounded interval except for the case mentioned.)

Inequality (1.21) restricts the values of the norm of $y^{(k)}$ in terms of the norms of y and $y^{(n)}$. Do all values of $\|y^{(k)}\|$ subject to the constraint (1.21) occur? Note that $\|y^{(k)}\| = 0$ implies that y is a polynomial of degree $k - 1$. So $\|y^{(k)}\|$ assumes the value zero if and only if J is bounded or J is unbounded and $p = \infty$ and $k = 1$. The next result shows that all values strictly between 0 and μ actually occur.

Let

$$Q(y) = \|y^{(k)}\| \left(\|y\| + \|y^{(n)}\| \right)^{-1} \tag{1.23}$$

for all

$$y \in D(Q) = W_{p,p}^n - \{0\}. \tag{1.24}$$

Clearly $\mu = \sup Q(y), y \in D(Q)$.

Remark 1.1 We will see in Section 3 that μ may or may not be in the range of Q. It is clear from (1.23) that Q assumes the value 0 only if (i) J is bounded (so that polynomials are in $L^p(J)$) or (ii) J is unbounded, $k = 1$, and $p = \infty$ in which case Q takes constant functions into 0.

Theorem 1.3 *Let p, J, n, k be as in Theorem 1.2. The range of Q contains the open interval $(0, \mu)$.*

Proof.

Case 1. J is unbounded.

Just as seen in the proof of Theorem 1.1, Q being continuous must have a connected range. Thus if $y, z \in D(Q)$ and $Q(y) = \alpha$, $Q(z) = \beta$, then the range of Q contains $[\alpha, \beta]$.

¿From the definition of μ it follows that β can be chosen arbitrarily close to μ. The proof is complete if we can show that there are arbitrarily small positive numbers in the range of Q. It is not difficult to treat the case $k = 1$ or $p = \infty$. So suppose $k > 1$, $p < \infty$. Consider a function y in $C^{(n)}(J)$ which is nonnegative, not identically zero, has compact support I in J and is a constant in a neighborhood of its maximum; i.e., the graph of y looks like this:

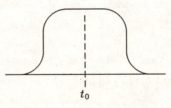

t_0

For any $h > 0$ let y_h be the function obtained from y by pulling its graph apart a distance h at the point $t_0, y(t_0)$):

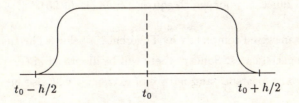

$t_0 - h/2$ t_0 $t_0 + h/2$

Then $\|y_h^{(i)}\| = \|y^{(i)}\|$ for $i = k, n$, $h > 0$ but $\|y_h\| \to \infty$ as $h \to \infty$. Hence $Q(y)$ can be made arbitrarily small by choosing h large enough and t_0 far enough away from the finite end point of J if J has a finite end point.

Case 2.   The interval J is bounded. In this case, to see that $Q(y)$ assumes arbitrarily small positive values, we need only observe that y can be replaced by $y + c$ for any constant c. This leaves $y^{(k)}$, $y^{(n)}$ unchanged and makes $\|y + c\|$ arbitrarily large by choosing c large enough. $\square$

1.3   Inequalities of Product Form

Here we consider the inequality

$$\|y^{(k)}\|_q \leq K \, \|y\|_p^\alpha \, \|y^{(n)}\|_r^\beta \ . \tag{1.25}$$

Let $\dot{W}_{p,r}^n(J)$ denote all y in $W_{p,r}^n(J)$ whose support is contained in a compact subinterval (which may be different for different y) of the interior of J.

The simple example $n = 2$, $k = 1$, and $y' = 1$ shows that (1.25) does not hold on bounded intervals J. In case of an unbounded interval J we say that inequality (1.25) is valid whenever there is a positive number K such that (1.25) holds for all y in $W_{p,r}^n(J)$. Three basic questions about (1.25) are:

Question 1.   Given an unbounded interval J, positive integers k, n with $1 \leq k < n$ and numbers p, r satisfying $1 \leq p, r \leq \infty$, for what values of the parameters α, β and q is (1.25) valid?

Question 2.   Given that (1.25) is valid, there clearly is a smallest constant K. This best constant is denoted by $K = K(n, k, p, q, r, J)$ to highlight its dependence on these quantities. (Theorem 1.4 below will show that α and β are uniquely determined by n, k, p, q and r.) What are the exact values of the constants $K(n, k, p, q, r, J)$ for $J = R$ or $J = R^+$?

Question 3.   Suppose (1.25) is valid and $K = K(n, k, p, q, r, J)$. Are there extremals? An extremal is a nontrivial function y for which equality holds in (1.25).

The first question is answered completely by Theorem 1.4 below. The answer to question 2 is known only in a few special cases. Some of these will be discussed in Chapter 2. Extremals may or may not exist. The existence and nature of extremals will also be further discussed in Chapter 2.

Theorem 1.4 *(Gabushin [1967].) Let $J = (a, \infty)$, $-\infty \leq a < \infty$. Let k, n be integers with $0 \leq k < n$. Let $1 \leq p, r \leq \infty$. There is a positive constant K such that inequality (1.25)* *holds for all $y \in W_{p,r}^n(J)$ if and only if*

$$nq^{-1} \leq (n-k)p^{-1} + kr^{-1} \tag{1.26}$$

and

$$\alpha = (n - k - r^{-1} + q^{-1})/(n - r^{-1} + p^{-1}), \quad \beta = 1 - \alpha. \tag{1.27}$$

Gabushin's Theorem is closely related to a theorem due to Nirenberg [1959] that extends the classical Sobolev inequality. The difference is that no norms of the intermediate derivatives of y are needed on the right-hand side of the inequality (1.25) here and that Gabushin's inequality only holds for unbounded J, but is not restricted to compact support functions y.

Before presenting a proof we make some observations. Given p, r, n, k satisfying the conditions of the theorem, if q satisfies (1.26) then (1.25) holds for some constant K provided α, β are determined by (1.27). We follow the usual convention that $q^{-1} = 0$ when $q = \infty$. It follows from the change of variable $t \to t + a$ and $t \to -t$ that $K(n, k, p, q, r, J) = K(n, k, p, q, r, R^+)$ for any $J = (a, \infty)$ or $J = (-\infty, a)$ with $-\infty < a < \infty$. On the other hand, as we will see later, these constants are not the same in general for $J = R$ and $J = R^+$. Thus there are only two distinct cases to consider: $J = R$ and $J = R^+$.

In the special case $p = r$, (1.27) reduces to $q \geq p$. If $p = r = \infty$ then only $q = \infty$ satisfies (1.26). For $p = \infty$ and r finite (1.27) holds when $nrk^{-1} \leq q \leq \infty$. Similarly when $r = \infty$ and p finite we must have $np/(n-k) \leq p \leq \infty$.

Proof. (of Theorem 1.4) The necessity part of the proof is taken from Gabushin [1967]. Our sufficiency proof, although somewhat related to Gabushin's, has some new features.

(a) Necessity. Suppose (1.25) holds for all $y \in W_{p,r}^n(J)$, $J = R$ or R^+ with some fixed constant K. Choose a nontrivial C^∞ function ϕ with compact support in $(0,1)$. Consider the class of functions

$$\phi_{m,s,t}(x) = s \sum_{j=0}^{m-1} \phi(tx - j/t)$$

with s, t any positive numbers and m any positive integer. Clearly ϕ and each of $\phi_{m,s,t}$ is in $W_{p,r}^n(J)$. Hence (1.25) holds for each $\phi_{m,s,t}$. A computation yields that

$$\|\phi_{m,s,t}^{(i)}\|_p = m^{1/p} s t^{i-1/p} \|\phi^{(i)}\|_p, i = 0, \ldots, n. \tag{1.28}$$

By substituting (1.28) into (1.25) we get

$$m^{1/q} s t^{k-1/q} \|\phi^{(k)}\|_q \leq K m^{\alpha/p + \beta/r} s^{\alpha+\beta} t^{-\alpha/p + \beta(n-1/r)} \|\phi\|_p^\alpha \|\phi^{(n)}\|_r^\beta. \tag{1.29}$$

Since we can let s and t go to zero and to $+\infty$ and m can be an arbitrarily large positive integer it follows that

(i) $\alpha + \beta = 1$

(ii) $k - q^{-1} = -\alpha p^{-1} + \beta(n - r^{-1})$

(iii) $q^{-1} \leq \alpha p^{-1} + \beta r^{-1}$

and (1.26), (1.27) follow.

(b) Sufficiency. This half of the proof of Theorem 1.4 is established with the help of several lemmas, some of which are of independent interest. □

Lemma 1.4 *Suppose $J = [a, b]$ is a closed and bounded interval. If f is a differentiable function on J satisfying $f'(t) \geq c > 0$ or $-f'(t) \geq c > 0$ for all t in J, then there exists a subinterval $[a_1, b_1]$ of J of length $(b-a)/4$ such that $|f(t)| \geq c(b-a)/4$, $t \in [a_1, b_1]$.*

Proof. Suppose $f'(t) \geq c > 0$ for t in J. Then f has at most one zero in J. If f has no zero in J then either f or $-f$ is positive on J. If f is positive then

$$f(t) \geq f(a) + c(t - a), \quad t \in J$$

and hence

$$f(t) \geq c(b-a)/4, \quad t \in [a + (b-a)/4, b].$$

The case when $-f$ is positive is similar. The case when f has a zero at d can be reduced to the above by replacing the interval $[a, b]$ with $[a, d]$ or $[d, b]$, whichever has length $\geq (b-a)/2$. □

In general, knowing that y is in $L^p(a, \infty)$ does not provide any information about the pointwise asymptotic behavior of $y(t)$ as $t \to \infty$. However, knowing that y and some derivative of y are in $L^p(a, \infty)$ does provide such information.

Lemma 1.5 *If $y \in W_{p,r}^n(R^+)$, $1 \leq p < \infty$, $1 \leq r \leq \infty$, n a positive integer, then*

$$y^{(k)}(t) \to 0 \text{ as } t \to \infty, \quad k = 0, 1, \ldots, n-1. \tag{1.30}$$

If $p = \infty$, $r < \infty$, and $n > 1$, then (1.30) holds for $k = 1, \ldots, n-1$.

Proof. If $p = r$, the lemma follows immediately from Theorem 1.2. The general case can also be established if Theorem 1.2 can be duly generalized. We give here a direct proof of the lemma.

Case 1. $p < \infty, r < \infty.$

Suppose $y \in L^p(0, \infty)$, $y^{(n)} \in L^r(0, \infty)$ and (1.30) does not hold for $k = n - 1$. Then there exists a $K > 0$, and a sequence $t_m \to \infty$ such that either $y^{(n-1)}(t_m) > 2K$ or $y^{(n-1)}(t_m) < -2K$. We may assume the former since in the latter case we can replace y by $-y$. For $t \in I_j = [t_j, t_j + 1]$ we get from Hölder's inequality

$$
\begin{aligned}
y^{(n-1)}(t) &= y^{(n+1)}(t_j) - \int_t^{t_j} y^{(n)} \\
&\geq 2K - \|y^{(n)}\|_{r,j} \to 2K \text{ as } j \to \infty.
\end{aligned}
\tag{1.31}
$$

Here $\|\cdot\|_{r,j}$ denotes the usual L^r norm on the interval I_j. For all sufficiently large j, $y^{(n-1)}(t) \geq K$ for $t \in I_j$. If $n = 1$, this contradicts $y \in L^p(R^+)$ and the proof is complete. If $n > 1$, we may conclude from Lemma 1.4 that $y^{(n-2)}(t) > K/4$ in some subinterval J_j of I_j of length $1/4$. Repeated applications of Lemma 1.4 yield that $y(t) \geq K/4^h$, $h = 1 + 2 + \ldots + n - 1$ for t in some subinterval of I_j of length 4^{1-n}. This contradicts $y \in L^p(R^+)$. Having proved that $y^{(n-1)}(t) \to 0$ as $t \to \infty$ we can use the same technique to show that $y^{(n-2)}(t) \to 0$ as $t \to \infty$. By repeating this argument we arrive at (1.30).

Case 2. $1 \leq p < \infty, r = \infty.$

Here we proceed as in Case 1, but use the intervals $I_j = [t_j, t_j + K/\|y^{(n)}\|_{\infty,j}]$. In place of (1.31) we have

$$
y^{(n-1)}(t) = y^{(n-1)}(t_j) - \int_t^{t_j} y^{(n)} \geq 2K - K = K.
$$

Case 3. $p = \infty, 1 \leq r < \infty.$

Again we use the same argument as in Case 1 with the intervals

$$
I_j = [t_j, t_j + a_j] \text{ where } a_j = \|y^{(n)}\|_{r,j}^{-1}.
$$

In place of (1.31) we have

$$
\begin{aligned}
y^{(n-1)}(t) &= y^{(n-1)}(t_j) - \int_t^{t_j} y^{(n)} \\
&\leq 2K - \|y^{(n)}\|_{r,j} a_j^{1/r'} = 2K - \|y^{(n)}\|_{r/j}^{1/r} \to 2K \text{ as } j \to \infty.
\end{aligned}
$$

Hence $y^{(n-1)}(t) \geq K$ for all $t \in I_j$ and all j sufficiently large. ¿From Lemma 1.4 we get $y'(t) \geq K4^{-h}$, $h = 1 + 2 + \ldots + n - 2$ for t in some subinterval of I_j of length $4^{1-n}a_j \to \infty$ as $j \to \infty$. This contradicts the boundedness of y and completes the proof of Lemma 1.5. $\square$

Next we show that functions in $W_{p,r}^n(R^+)$ can be approximated by functions of compact support when not both of p and r are infinite.

Lemma 1.6 *Let $y \in W_{p,r}^n(R^+)$ with not both of p, r infinite and when $p = \infty$ either $r > 1$ or $n > 1$. Given any $\epsilon > 0$ there is some $T > 0$ and a function $y_0 \in \dot{W}_{p,r}^n(R^+)$ such that*

$$y_0(t) = y(t) \text{ for } t \in (0, T]$$

and

$$\|y - y_0\|_p < \epsilon \text{ if } p < \infty \tag{1.32}$$

$$\|y_0\|_\infty \leq \|y\|_\infty \text{ if } p = \infty \tag{1.33}$$

$$\|y^{(k)} - y_0^{(k)}\|_r < \epsilon \text{ if } r < \infty, \quad k = 1, \ldots, n \tag{1.34}$$

$$\|y_0^{(k)}\|_\infty \leq \|y^{(k)}\|_\infty + \epsilon \text{ if } r = \infty, \quad k = 1, \ldots, n. \tag{1.35}$$

Proof. First consider the case $1 \leq p < \infty$. Let ϕ be a C^∞ function on $[0,1]$ which is 1 in a right neighborhood of 0, 0 in a left neighborhood of 1, and satisfies $0 \leq \phi(t) \leq 1$. For each positive integer j define

$$\psi_j(t) = \begin{cases} 1, & 0 \leq t \leq j \\ \phi(t - j), & j \leq t \leq j + 1 \\ 0, & j + 1 \leq t \end{cases} \tag{1.36}$$

and consider

$$y_j = y\psi_j. \tag{1.37}$$

Then $y_j = y$ on $[0, j]$. Since $0 \leq \psi_j \leq 1$ we have

$$\|y - y_j\|_p^p = \int_j^\infty |y - y_j|^p \leq \int_j^\infty |y|^p \to 0 \text{ as } j \to \infty. \tag{1.38}$$

From Leibnitz's formula for the derivatives of a product we get

$$y_j^{(k)} = \sum_{i=0}^k \binom{k}{i} y^{(k-i)} \psi_j^{(i)}, \quad k = 1, \ldots, n.$$

Hence when $r < \infty$ we get

$$\|y^{(k)} - y_j^{(k)}\|_r \leq \|y^{(k)} - y^{(k)}\psi_j\|_r + \sum_{i=1}^k \binom{k}{i} \|y^{(k-i)} \psi_j^{(i)}\|_j. \tag{1.39}$$

The first term on the right goes to zero as $j \to \infty$, by an argument similar to that in (1.38). The convergence to zero as $j \to \infty$ of the second term on the right in (1.39) follows from the uniform boundedness of $\psi_j^{(i)}$, $i = 1, \ldots, n$, the fact that the support of $y^{(n-i)} \psi_j^{(i)}$ is contained in $(j, j + 1)$, $i = 1, \ldots, n$ and (1.30) of Lemma 1.5.

For $r = \infty$ and $\epsilon > 0$ we have

$$|y_j^{(k)}(t)| \le |y^{(k)}(t)\psi_j(t)| + \left\| \sum_{i=1}^{k} \binom{k}{i} y^{(k-i)} \psi_j^{(i)} \right\|_\infty$$

$$\le |y^{(k)}(t)| + \epsilon,$$

for j sufficiently large by (1.30) and (1.36). Hence

$$\|y_j^{(k)}\|_\infty \le \|y^{(k)}\|_\infty + \epsilon,$$

for j sufficiently large, $k = 1, \ldots, n$.

In case $p = \infty$ we define ψ_j by

$$\psi_j(t) = \begin{cases} 1, & 0 \le t \le j \\ \phi((t-j)/j), & j \le t \le 2j \\ 0, & 2j \le t. \end{cases} \tag{1.40}$$

Then

$$\|\psi_j^{(k)}\|_r = j^{-k+1/r} \|\phi^{(k)}\|_r \to 0 \text{ as } j \to \infty$$

for $k = 1, 2, \ldots, n$ except when $r = 1$. We define y_j by (1.37) using (1.40) and proceed as before to get (1.33) with $y_0 = y_{\bar{j}}$ for j large enough.

Choosing $y_0 = y_j$ for j large enough we get (1.32), (1.33), (1.34), (1.35) for each case. $\square$

Lemma 1.7 *If $y \in W_{p,r}^1([a,b])$, $1 \le p, r \le \infty$, $b - a = 1$, then*

$$\|y\|_\infty \le \|y\|_p + \|y'\|_r. \tag{1.41}$$

Proof. Let $|y|$ attain its minimum at $t_1 \in [a,b]$. Then $|y(t_1)| \le \|y\|_p$. For any $t \in [a,b]$,

$$|y(t)| \le |y(t_1)| + \left| \int_{t_1}^t y' \right|$$

$$\le \|y\|_p + |t - t_1|^{1/r'} \|y'\|_r$$

$$\le \|y\|_p + \|y'\|_r$$

and (1.41) follows. Here $r'^{-1} + r^{-1} = 1$. $\square$

Lemma 1.8 *Let $y \in W_{p,r}^1([a,b])$, a,b finite, $1 \le p, r \le \infty$. Suppose $y(c) = 0$ for some c in $[a,b]$. Then*

$$\|y\|_\infty \le \|y\|_p + \|y'\|_r.$$

Here all norms are on $[a,b]$.

Proof. Let $[a, b] = \cup_{k=0}^{n} I_k$ where the intervals I_k have no interior point in common for $k = 1, \ldots, n$; c is in I_0; the length of I_k is 1 for $k = 1, \ldots, n$; and the length of $I_0 \leq 1$. Then for $t \in I_k$, by Lemma 1.7

$$\begin{aligned} |y(t)| &\leq \|y\|_{p,I_k} + \|y'\|_{r,I_k} \\ &\leq \|y\|_{p,[a,b]} + \|y'\|_{r,[a,b]}, \quad k = 1, \ldots, n \end{aligned}$$

and for $t \in I_0$

$$\begin{aligned} |y(t)| &\leq |y(c)| + \left| \int_c^t y' \right| \leq |t - c|^{1/r'} \|y'\|_{r,I_0} \\ &\leq \|y'\|_{r,[a,b]}, \end{aligned}$$

where $r'^{-1} + r^{-1} = 1$. The conclusion follows. $\square$

Lemma 1.9 *Let $1 \leq p, r \leq \infty$. Suppose $q \geq p$ and*

$$\alpha = (1 - r^{-1} + q^{-1})/(1 - r^{-1} + p^{-1}), \quad \beta = 1 - \alpha, \tag{1.42}$$

where $r^{-1} = 0$ if $r = \infty$. Then there exists a constant K such that for any $y \in W_{p,r}^1([a, b])$, $-\infty < a < b < \infty$, which has at least one zero in $[a, b]$, we have

$$\|y\|_q \leq K \, \|y\|_p^\alpha \, \|y'\|_r^\beta. \tag{1.43}$$

Proof. Let $\lambda > 0$ and consider $f(t) = y(\lambda t)$ on $[a/\lambda, b/\lambda]$. Then

$$\|f\|_\infty = \|y\|_\infty, \quad \|f\|_p = \lambda^{-1/p}\|y\|_p, \quad \|f'\|_r = \lambda^{1-1/r}\|y'\|_r, \tag{1.44}$$

where each norm is over the appropriate interval on which the function is defined. Applying Lemma 1.8 to f we get

$$\|y\|_\infty \leq \lambda^{-1/p}\|y\|_p + \lambda^{1-1/r}\|y'\|_r, \quad \lambda > 0. \tag{1.45}$$

Choosing $\lambda = (\|y\|_p/\|y'\|_r)^{p(1-\alpha)}$ yields (1.43) for $q = \infty$. (Note that $\|y'\|_r \neq 0$ unless $y \equiv 0$.) For $q < \infty$ we observe that

$$\int |y|^q = \int |y|^p |y|^{q-p} \leq \|y\|_\infty^{q-p}\|y\|_p^p$$

and estimate $\|y\|_\infty$ using (1.45). This completes the proof. $\square$

Proof. (of Theorem 1.4) We now return to the sufficiency proof of Theorem 1.4. First some simplifications. It suffices to prove (1.25) for $J = R^+$. The case $J = R$ will follow from Theorem 1.5 below or by applying the inequality to $(-\infty, 0)$ and $(0, \infty)$ and then combining the two resulting inequalities.

Case 1. $p = \infty$, $n = 1$. In this exceptional case we must have $k = 0$, $q = \infty$, $\alpha = 1$, $\beta = 0$ and so (1.25) is satisfied trivially.

Case 2. $p = \infty$, $r = \infty$. By (1.26), (1.27) we have $q = \infty$, $\alpha = (n - k)/n$, $\beta = k/n$. It suffices to establish the case $n = 2$, $k = 1$. The case $k = 0$ is trivial and the case $n > 2$ follows by an induction argument, which is described in detail under case 3.

Let $y \in W^2_{\infty,\infty}(R^+)$. By Taylor's formula

$$y(t + s) = y(t) + sy'(t) + \int_0^t (s - u)y''(t + u)du$$

or

$$y'(t) = s^{-1}[y(t + s) - y(t)] - s^{-1}\int_0^t (s - u)y''(t + u)du \text{ for } s, t > 0. \tag{1.46}$$

The integral in (1.46) is dominated by $s^2\|y''\|_\infty/2$. Hence

$$\|y'\|_\infty \leq 2s^{-1}\|y\|_\infty + (s/2)\|y''\|_\infty. \tag{1.47}$$

Now minimizing the right hand side over s in $(0, \infty)$ we find that the minimum value is attained when

$$s = 2 \|y\|^{1/2} \|y''\|^{-1/2}.$$

This gives

$$\|y'\|^2_\infty \leq 4 \|y\|_\infty \|y''\|_\infty. \tag{1.48}$$

Case 3. Either $1 \leq p < \infty$ or $1 \leq r < \infty$. First we show that we may restrict consideration to functions whose supports lie in some interval $[0, T]$. Let $y \in W^n_{p,r}(R^+)$ and $\epsilon > 0$. By Lemma 1.6 there is some $T > 0$ and $y_0 \in W^n_{p,r}(R^+)$ such that $y_0 = y$ on $(0, T]$, $y_0 = 0$ on $[T + 1, \infty)$ and (1.32) to (1.35) hold. From

$$\|y_0^{(k)}\|_q \leq K \|y_0\|_p^\alpha \|y_0^{(n)}\|_r^\beta \tag{1.49}$$

it follows that

$$\|y^{(k)}\|_{q,[0,T]} \leq \|y_0^{(k)}\|_q \leq K (\|y\|_p + \epsilon)^\alpha (\|y^{(n)}\|_r + \epsilon)^\beta.$$

Letting $\epsilon \to 0$ and $T \to \infty$ yields (1.25). Thus if (1.25) holds for functions in $W^n_{p,r}(R^+)$ which are supported on intervals $(0, T)$, $T > 0$ then (1.25) holds for all $y \in W^n_{p,r}(R^+)$.

The proof is by induction on n. Let $y \in W^n_{p,r}(R^+)$. Suppose $n = 1$, $k = 0$. Define y_0 as in the proof of Lemma 1.6, i.e., $y_0 = y_j$ where y_j is given by (1.37) with ψ_j defined by (1.36) for $p < \infty$ and by (1.40) when $p = \infty$. Then $y_0 = y$ on $(0, j)$ and $y_0(j + 1) = 0$. Also $q > p$ by (1.26) and α, β are given by (1.42). Thus (1.49) holds by Lemma 1.9. Hence (1.25) holds by the above approximation argument. Now suppose $y \in W_{p,r}^n(R^+)$, $1 \le p$, $r \le \infty$ with at least one of p, r finite, $n = 2$, $k = 1$, and y is supported on some interval $(0, T)$, $T < \infty$. Then $y^{(k)} \in L^q(R^+)$ for all k, $0 \le k < n$ and all q, $1 \le q \le \infty$. Define

$$S = \{t > 0 \mid y'(t) \ne 0\}.$$

Then S is an open set in the relative topology of R^+. Hence $S = \cup_{n=1}^\infty I_n$, where the I_n's are disjoint and each I_n is an open interval in the relative topology of R^+. In each interval I_n, y' is of constant sign and vanishes at at least one end point. We first show that inequality (1.25) holds if the norms are interpreted as being taken over I_n. We may assume that $y' > 0$ in I_n; otherwise, we replace y by $-y$. Then y is increasing. If y has a zero in $I_n = [a, b]$, then from (1.43) with $q = \infty$ we have

$$y(b) - y(a) \le 2 \, \|y\|_\infty \le 2K \, \|y\|_p^{\alpha_1} \, \|y'\|_r^{\beta_1}.$$

If y has no zero in I_n, then (1.43) applied to $y - y(a)$ if $y(a) > 0$ or to $y(b) - y$ if $y(a) < 0$ yields

$$
\begin{aligned}
y(b) - y(a) &= \|y - y(a)\|_\infty \text{ (or } \|y(b) - y\|_\infty) \\
&\le K \, \|y - y(a)\|_p^{\alpha_1} \, \|y'\|_r^{\beta_1}.
\end{aligned}
$$

In either case

$$y(b) - y(a) \le 2K \, \|y\|_p^{\alpha_1} \, \|y'\|_r^{\beta_1}. \tag{1.50}$$

Applying (1.43) to y' yields

$$\|y'\|_q \le K \, \|y'\|_1^{\alpha_2} \, \|y''\|_r^{\beta_2}. \tag{1.51}$$

Noting that $\|y'\|_1 = y(b) - y(a)$, (1.50) and (1.51) yield (1.25). (Observe that α (and hence β) in (1.50) and (1.51) are not the same since $q = \infty$ in (1.50) and $q < \infty$ in (1.51). See Lemma 1.9 for the definition of α.) To obtain inequality (1.25) on R^+ we first consider the two extreme cases: $q = \infty$ and $q = s$ where $2s^{-1} = p^{-1} + r^{-1}$. In the former case

$$
\begin{aligned}
\|y'\|_\infty &= \sup_n \{\|y'\|_{\infty, I_n}\} \\
&\le \sup_n \{K \, \|y\|_{p, I_n}^\alpha\} \, \|y''\|_{r, I_n}^{1-\alpha} \\
&\le K \, \|y\|_p^\alpha \, \|y''\|_r^{1-\alpha}.
\end{aligned}
\tag{1.52}
$$

In the last step we used the fact that the constant K in (1.51) is independent of the interval I_n.

In the case $q = s$, $\alpha = 1/2$ and

$$\|y'\|_s^s = \sum_{n=1}^\infty \|y'\|_{s, I_n}^s$$

$$\leq \quad K^s \sum_{n=1}^{\infty} (\|y\|_{p,I_n}^p)^{s/2p} \, (\|y''\|_{r,I_n}^r)^{s/2r}$$

$$\leq \quad K^s \left(\sum_{n=1}^{\infty} \|y\|_{p,I_n}^p \right)^{s/2p} \left(\sum_{n=1}^{\infty} \|y''\|_{r,I_n}^r \right)^{s/2r}$$

$$= \quad K^s \, \|y\|_p^{s/2} \, \|y''\|_r^{s/2}. \tag{1.53}$$

The last inequality follows from the discrete Hölder inequality.

For $s < q < \infty$, inequality (1.25) follows from (1.52), (1.53) and

$$\int |y'|^q = \int |y'|^s \, |y'|^{q-s} \leq \|y'\|_\infty^{q-s} \, \|y'\|_s^s.$$

Finally we establish (1.25) for functions y in $\dot{W}_{p,r}^n(R^+)$ by induction on n. (It is here that we use the fact that y has compact support to ensure that y and its derivatives up to order $n-1$ are in $L^q(R^+)$ for all q, $1 \leq q \leq \infty$.) We have just proved the case $n = 2$. Suppose (1.25) holds for $n-1$ and all k, $0 \leq k < n-1$. It suffices to prove only the case $k = n-1$ since the cases $k < n-1$ follow from the inductive hypothesis and this case. Let s be a real number such that $(n-1)s^{-1} = (n-2)p^{-1} + q^{-1}$. Then $(n-1)q^{-1} \leq (n-2)r^{-1} + s^{-1}$. By (1.25) with $y \in \dot{W}_{p,q}^{n-1}$ and $p, s, q, n-1, 2$ in place of p, q, r, n, k we get

$$\|y'\|_s \leq K_1 \, \|y\|_p^{\alpha_1} \, \|y^{(n-1)}\|_q^{\beta_1}. \tag{1.54}$$

By (1.25) applied to $y' \in \dot{W}_{s,r}^{n-1}$ with $s, q, r, n, n-1$ in place of p, q, r, n, k we get

$$\|y^{(n-1)}\|_q \leq K_2 \, \|y'\|_s^{\alpha_2} \, \|y^{(n)}\|_r^{\beta_2}. \tag{1.55}$$

In (1.54) and (1.55) $\alpha_1, \alpha_2, \beta_1, \beta_2$ are chosen according to (1.26). Now (1.25) follows upon using (1.54) in (1.55). The proof of Theorem 1.4 is complete. $\square$

Clearly inequality (1.25) cannot hold on a finite interval J since for $y^{(k)} = 1$ and $y^{(n)} = 0$, the right-hand side is zero and the lefthand side is not. However, note that the proof of Theorem 1.4 contains the following.

Corollary 1.1 *Suppose J is a bounded interval, $1 \leq p$, $r \leq \infty$, and n, k are integers with $0 \leq k < n$. Then there exists a constant K such that for all $y \in W_{p,r}^n(J)$ satisfying $y^{(k)}(c_k) = 0$ for some $c_k \in J$, $k = 1, \ldots, n-1$ we have*

$$\|y^{(k)}\|_q \leq K \, \|y\|_p^\alpha \, \|y^{(n)}\|_r^\beta,$$

where α, β, q satisfy (1.26) and (1.27) and the norms are taken over J.

Theorem 1.5 *Let $J = R$ or R^+, $1 \leq p$, $r \leq \infty$, n, k positive integers with $k < n$. Then the following statements are all equivalent:*

1. $nq^{-1} \leq (n-k)p^{-1} + kr^{-1}$ and $\alpha = (n - k - r^{-1} + q^{-1})/(n - r^{-1} + p^{-1})$, $\beta = 1 - \alpha$.

2. There exists a constant K such that

$$\|y^{(k)}\|_q \leq K \, \|y\|_p^{\alpha} \, \|y^{(n)}\|_r^{\beta} \tag{1.56}$$

for all $y \in W_{p,r}^n(J)$.

3. There exists a constant K such that for all $\lambda > 0$ and all $y \in W_{p,r}^n(J)$ we have

$$\|y^{(k)}\|_q \leq K \, (\lambda^a \|y\|_p + \lambda^b \|y^{(n)}\|_r), \tag{1.57}$$

where $a\alpha + b\beta = 0$.

4. There exists a constant K such that for all $y \in W_{p,r}^n(J)$

$$\|y^{(k)}\|_q \leq K(\|y\|_p + \|y^{(n)}\|_r). \tag{1.58}$$

5. If $y \in W_{p,r}^n(J)$ then

$$y^{(k)} \in L^q(J), \quad k = 1, \ldots, n - 1. \tag{1.59}$$

The constant K need not be the same in these statements but in each case there is a smallest constant K which depends on p, q, r, n, k and J but not on y and in the case of (1.57) K does not depend on λ.

Proof. The equivalence of (1) and (2) was established in Theorem 1.4. Inequality (1.57) follows from (1.25) and the general inequality between weighted arithmetic and geometric means (Hardy-Littlewood and Polya [1934]):

$$\begin{aligned} \|y^{(k)}\|_q &\leq K \, \|y\|_p^{\alpha} \, \|y^{(n)}\|_r^{\beta} \\ &= K \, (\lambda^a \|y\|_p)^{\alpha} \, (\lambda^b \|y^{(k)}\|_r)^{\beta} \\ &\leq K \, (\alpha\lambda^a \|y\|_p + \beta\lambda^b \|y^{(n)}\|_r). \end{aligned}$$

Inequality (1.58) is a special case of (1.57) and statement (5) clearly follows from (4). Thus we have shown that $(1) \Leftrightarrow (2) \Rightarrow (3) \Rightarrow (4) \Rightarrow (5)$.

To show that $(5) \Rightarrow (4)$ we use the closed graph theorem. The linear space $W_{p,r}^n(J)$ can be made into a Banach space with the norm

$$\|y\| = \|y\|_p + \|y^{(n)}\|_r. \tag{1.60}$$

The differentiation operator $A = d^k/dt^k$ is a closed linear operator from $W_{p,r}^n(J)$ into $L^q(J)$. By (5) A is defined on all of $W_{p,r}^n(J)$. Hence A is bounded by the closed graph theorem, i.e., (1.58) holds.

To show that $W_{p,r}^n(J)$ is a Banach space with the norm (1.60), we consider the notion of a weak derivative. A locally integrable function y is n times weakly differentiable in $L^r(J)$ if there exists a function z in $L^r(J)$ such that

$$\int_J y\phi^{(n)} = (-1)^n \int_J z\phi$$

for all ϕ infinitely differentiable with compact support in the interior of J. Such a z is a weak n^{th} derivative of y. It is unique a.e.. Sobolev's lemma, see Adams [1975], in the case of functions of one variable, states that weak differentiability implies strong differentiability, i.e., there exists a function w such that $y = w$ a.e. and w has a classical n^{th} derivative. (Although Sobolev's lemma is usually stated only for functions y such that $y, y', \ldots, y^{(n)}$ all belong to L^p, it applies in our situation since the restrictions to compact subintervals will be in L^1.)

To show that the space is complete, we let y_k be a Cauchy sequence in $W_{p,r}^n(J)$ in the norm (1.49). Then y_k and $y_k^{(n)}$ are Cauchy sequences in $L^p(J)$ and $L^r(J)$, respectively. Let $y_k \to y$ in $L^p(J)$ and $y_k^{(n)} \to z \in L^r(J)$. For any C_0^∞ function ϕ we have

$$\int_J y\phi^{(n)} = \lim_{n\to\infty} \int_J y_k\phi^{(n)} = (-1)^n \lim_{n\to\infty} \int_J y_k^{(n)}\phi = (-1)^n \int_J z\phi.$$

Thus z is the weak n^{th} derivative of y. Hence there exists a classically n times differentiable function $w = y$ a.e. such that $w^{(n)} = z$ a.e. Thus $w \in W_{p,r}^n(J)$ and is the limit of the Cauchy sequence y_k.

The proof that the linear operator $Ay = y^{(k)}$ from $W_{p,r}^n(J)$ into $L^q(J)$ is a closed operator is similar. This completes the proof of the implication $(5) \Rightarrow (4)$.

To show that (4) implies (3) let

$$f(t) = y(\lambda t), \quad \lambda > 0.$$

Then $f \in W_{p,r}^n$ and

$$\|f^{(i)}\|_q^q = \lambda^{i-1} \int_J \left|y^{(i)}\right|^q (\lambda t)d(\lambda t) = \lambda^{i-1}\|y^{(i)}\|_q^q, \quad 1 \le q < \infty.$$

Applying (1.58) to f we get

$$\lambda^{(k-1)q}\|y^{(k)}\|_q \le K(\lambda^{-1/p}\|y\|_p + \lambda^{(n-1)/r}\|y^{(n)}\|_r)$$

which yields (1.57) with

$$a = -p^{-1} - kq^{-1} + q^{-1}, \quad b = nr^{-1} - r^{-1} - kq^{-1} + q^{-1}. \tag{1.61}$$

It is sufficient to establish (1.57) for one particular choice of a and b with $a \ne 0$ and $b \ne 0$; the general case can then be obtained by applying (1.57) to a power of λ. Using (1.26) a direct computation shows that $a\alpha + b\beta = 0$. This establishes (1.57) for p, q, r all finite. The case when at least one of p, q, r is infinite is verified similarly. The implication $(3) \Rightarrow (2)$ is established by considering the right-hand side of (1.57) as a function of λ, say $g(\lambda)$, and then minimizing g over all λ in $(0, \infty)$. This completes the proof of Theorem 2. $\square$

Corollary 1.2 *Suppose the hypotheses and the conditions of statement (1) of Theorem 1.5 hold. In addition, assume that $1 \leq q < \infty$ if $r = 1$ and $k = n - 1$. Then*

(i) Given $\epsilon > 0$ there exists a $K = K(\epsilon) > 0$ such that for all $y \in W_{p,r}^n(J)$ we have

$$\|y^{(k)}\|_q \leq \epsilon \|y\|_p + K(\epsilon) \|y^{(n)}\|_r.$$

(ii) Given $\epsilon > 0$ there exists a $K = K(\epsilon) > 0$ such that for all $y \in W_{p,r}^n(J)$

$$\|y^{(k)}\|_q \leq K(\epsilon) \|y\|_p + \epsilon \|y^{(n)}\|_r.$$

Proof. This follows from part (3) of Theorem 1.5. Note that $\alpha > 0$ and choose $a = \alpha^{-1}$, $b = -\beta^{-1}$. Then, for part (i), choose λ so small that $K\lambda^a < \epsilon$. For part (ii) choose λ so large that $K\lambda^b < \epsilon$. $\square$

Corollary 1.3 *Let $1 \leq p, r \leq \infty$, $J = R$ or R^+ and let n and m be positive integers. Let k_j be an integer satisfying $1 \leq k_j < n$ and suppose each q_j satisfies*

$$nq_j^{-1} \leq (n - k_j)p^{-1} + k_j r^{-1}, \quad j = 1, \ldots, m.$$

For $Q = (p, q_1, \ldots, q_m, r)$ let W_Q^n denote the set of all $y \in L^p(J)$ such that $y^{(n-1)}$ exists and is locally absolutely continuous and $y^{(k_j)} \in L^{q_j}(J)$, $j = 1, \ldots, m$, $y^{(n)} \in L^r(J)$. Then $W_{p,r}^n(J) = W_Q^n(J)$ and the two norms

$$\|y\|_1 = \|y\|_p + \|y^{(n)}\|_r$$

$$\|y\|_2 = \|y\|_p + \sum_{j=1}^{m} \|y^{(k_j)}\|_{q_j} + \|y^{(n)}\|_r$$

are equivalent.

Proof. This is an immediate consequence of (1.58). $\square$

Corollary 1.4 *Let n be a positive integer, let $J = R$ or R^+, $1 \leq p \leq \infty$. Let $W_p^n(J)$ denote the set of all functions y such that $y \in L^p(J)$, $y^{(n-1)}$ is locally absolutely continuous and $y^{(k)} \in L^p(J)$, for $k = 1, \ldots, n$. Then $W_{p,p}^n(J) = W_p^n(J)$ and the two norms*

$$\|y\|_1 = \|y\|_p + \|y^{(n)}\|_p$$

$$\|y\|_2 = \|y\|_p + \sum_{k=1}^{n} \|y^{(k)}\|_p$$

are equivalent.

Proof. This is a special case of Corollary 1.2, i.e., $p = r$, $m = n - 1$, $k_j = j$, $q_j = p$, $j = 1, \ldots, n - 1$, $Q = (p, p, \ldots, p)$. $\square$

Theorem 1.6 *Suppose the hypotheses of Theorem 1.5 hold. Let K be the smallest constant in inequality (1.25). Assume that u_0, u_k, u_n are positive numbers satisfying*

$$u_k < K u_0^\alpha u_n^\beta.$$

Then there exists a function y in $W_{p,r}^n(J)$ such that

$$\|y\|_p = u_0, \quad \|y^{(k)}\|_q = u_k, \quad \|y^{(n)}\|_r = u_n.$$

Proof. Suppose $1 \le p < \infty$. Define

$$Q(y) = \|y^{(k)}\|_q / (\|y\|_p^\alpha \|y^{(n)}\|_r^\beta), \quad y \in W_{p,r}^n(J), \quad y \ne 0.$$

(Note that $\|y^{(n)}\|_r \ne 0$ since $\|y^{(n)}\|_r = 0$ implies $y^{(n)} \equiv 0$ and so y is a polynomial. Since J is an unbounded interval, the only polynomial in $L^p(J)$ is the zero polynomial.) Then

$$K = \sup Q(y), \tag{1.62}$$

where the sup is taken over all $y \ne 0$ in $W_{p,r}^n(J)$.

It remains to show that the range of Q contains $(0, K)$. The proof is similar to that of Theorem 1.3 and is thus omitted. $\square$

Remark 1.2 Kwong and Zettl [1980] showed that for $1 < p = q = r \le \infty$, $J = R^+$, $n = 2$, $k = 1$, so that $\alpha = \beta = 1/2$, there exists a function $y \in W_{p,p}^2(R^+)$ such that $y \ne 0$ and $Q(y) = K$. Such a function is called an extremal. It is also shown in Kwong and Zettl [1980] that for $J = R$ and for all p satisfying $1 < p < \infty$, such extremal functions do not exist. In the same paper it is also shown that in case $J = R^+$ and $p = 1$, extremal functions do not exist.

1.4   Growth at Infinity

In this section we extend the asymptotic estimates of Lemma 1.5.

Theorem 1.7 *Let n denote a positive integer, let f, g be positive nondecreasing functions on R^+. If y is an n times (weakly) differentiable function on R^+ such that*

$$fy \in L^p(R^+) \text{ and } gy^{(n)} \in L^r(R^+)$$

for some $p, r, 1 \leq p, r \leq \infty$, then

$$y^{(k)}(t) = o(f^{-\alpha}(t)g^{-\beta}(t)), \quad t \to \infty, \quad k = 0, 1, \ldots, n-1, \tag{1.63}$$

where $\alpha = (n - k - r^{-1})/(n - r^{-1} + p^{-1})$, $\beta = 1 - \alpha$, unless $p = r = \infty$ in which case we can only conclude that

$$y^{(k)}(t) = O(f^{-\alpha}(t)g^{-\beta}(t)), \quad t \to \infty, \quad k = 0, 1, \ldots, n-1. \tag{1.64}$$

Proof. Assume $1 \leq p, r < \infty$. Since f and g are nondecreasing we have

$$f^p(t) \int_t^\infty |y|^p \leq \int_t^\infty |fy|^p$$

$$g^r(t) \int_t^\infty |y^{(n)}|^r \leq \int_t^\infty |gy^{(n)}|^r.$$

Now apply inequality (1.25) to y restricted to the interval (t, ∞) with $q = \infty$ to obtain

$$
\begin{aligned}
|y^{(k)}(t)| &\leq \|y^{(k)}\|_{\infty, (t,\infty)} \\
&\leq K \left(\int_t^\infty |y|^p \right)^{\alpha/p} \left(\int_t^\infty |y^{(n)}|^r \right)^{\beta/r} \\
&\leq K f^{-\alpha}(t) g^{-\beta}(t) \left(\int_t^\infty |fy|^p \right)^{\alpha/p} \left(\int_t^\infty |gy^{(n)}|^r \right)^{\beta/r}.
\end{aligned}
\tag{1.65}
$$

The last two integrals $\to 0$ as $t \to \infty$. Here we have used the fact that the constant K in inequality (1.25) is the same for all half lines.

If $p = \infty$ and $1 \leq r < \infty$ note that

$$f(t)\|y\|_{\infty, (t,\infty)} \leq \|fy\|_{\infty, (t,\infty)}$$

and proceed as above. The proof is similar for $1 \leq p < \infty$ and $r = \infty$. If $p = \infty = r$ then in place of (1.65) we get

$$|y^{(k)}(t)| \leq K f^{-\alpha}(t) g^{-\beta}(t) \|fy\|_{\infty, (t,\infty)} \|gy^{(n)}\|_{\infty, (t,\infty)}. \tag{1.66}$$

Clearly the two norms on the right-hand side of (1.66) are nonincreasing functions of t and hence bounded as $t \to \infty$. This completes the proof of Theorem 1.7. $\square$

If we choose $f = g$ in Theorem 1.7, we have

Corollary 1.5 *Let f be a positive nondecreasing function on R^+, n a positive integer and $1 \leq p$, $r \leq \infty$. If y is an n times (weakly) differentiable function such that $fy \in L^p(R^+)$ and $fy^{(n)} \in L^r(R^+)$ then*

$$y^{(k)}(t) = o(1/f(t)), \quad t \to \infty, \quad k = 0, 1, \ldots, n-1,$$

unless $p = r = \infty$, in which case we can only conclude that

$$y^{(k)}(t) = O(1/f(t)), \quad t \to \infty, \quad k = 0, 1, \ldots, n-1.$$

The special case $p = q = r = \infty$, $n = 2$, $k = 1$ (so that $\alpha = 1/2 = \beta$) of inequality (1.25):

$$\|y'\|_\infty^2 \leq K \|y\|_\infty \|y''\|_\infty \tag{1.67}$$

is called Landau's inequality when $J = R^+$ and $K = 4$ and Hadamard's inequality when $J = R$ and $K = 2$. Our next result extends (1.67).

Theorem 1.8 *Let $0 \leq a < 1$, $J = R$, or $J = R^+$. Suppose $y \in L^\infty(J)$, y' is locally absolutely continuous, and $y''|y|^a \in L^\infty(J)$. Then $y' \in L^\infty(J)$ and*

$$\|y'\|_\infty^2 \leq K \|y\|_\infty^{1-a} \|y''|y|^a\|_\infty \tag{1.68}$$

with $K = 2/(1-a)$ when $J = R$ and $K = 4/(1-a)$ when $J = R^+$. $\tag{1.69}$

Remark 1.3 When $a = 0$, (1.68) and (1.69) reduce to (1.67) for both cases $J = R$ and $J = R^+$ with the same constant K. These constants are known to be best possible when $a = 0$. See the survey paper by Kwong and Zettl [1980b]. The proof given below involves improper integrals when $0 < a < 1$ but not when $a = 0$. Note also that Theorem 1.7 is an extension of the Landau and Hadamard inequalities since $y''|y|^a$ might be bounded for some a, $0 < a < 1$ without y'' being bounded.

Proof. First consider the case $J = R^+$. It suffices to show that for any $x_0 \in R^+$,

$$|y'(x_0)| \leq K\|y\|_\infty^{1-a}\|y''|y|^a\|_\infty. \tag{1.70}$$

Suppose first that y' has a zero in R^+. If $y'(x_0) = 0$ then (1.64) clearly holds. If $y'(x_0) \neq 0$ there must be a zero of y', say x_1, nearest to x_0. Assume $x_0 < x_1$; the case $x_1 < x_0$ is treated similarly. Note that y' is of constant sign, say positive on (x_0, x_1). (If y' is negative replace y by $-y$.) Now

$$
\begin{aligned}
|y'(x_0)|^2 &= |y'(x_1)|^2 - \int_{x_0}^{x_1} 2y'(x)y''(x)dx \\
&\leq 2\int_{x_0}^{x_1} y'(x)|y|^{-a}(x)|y''(x)||y|^a(x)dx \tag{1.71} \\
&\leq 2\|y''|y|^a\|_\infty \int_{x_0}^{x_1} y'(x)|y|^{-a}(x)dx.
\end{aligned}
$$

If y has no zero in $[x_0, x_1]$, then

$$\int_{x_0}^{x_1} y'(x)|y|^{-a}(x)dx = (1-a)^{-1}||y|^{1-a}(x_1) - |y|^{1-a}(x_0)|. \qquad (1.72)$$

If y has a zero x_2 in $[x_0, x_1]$, then the integral in (1.72) is improper. By dividing the interval of integration into $[x_0, x_2]$ and $[x_2, x_1]$ we see that

$$\int_{x_0}^{x_1} y'(x)|y|^{-a}(x)dx = (1-a)^{-1}[|y|^{1-a}(x_1) + |y|^{1-a}(x_0)]. \qquad (1.73)$$

In both cases we have

$$\int_{x_0}^{x_1} y'(x)|y|^{-a}(x)dx \le 2(1-a)^{-1}||y||_{\infty}^{1-a}. \qquad (1.74)$$

Thus (1.68) is established in case y' has a zero in R^+.

Suppose y' has no zero in R^+. Then y' is of constant sign and so we may assume y' is positive. By the mean value theorem there is a point x_n in $[0, n]$ such that

$$y'(x_n) = n^{-1}(y(n) - y(0)) \le 2n^{-1}||y||_{\infty} \to 0 \text{ as } n \to \infty.$$

Now by repeating the arguments above using x_n in place of x_1 inequality (1.71) becomes

$$\begin{aligned}
|y'(x_0)|^2 &\le |y'(x_n)|^2 + 2||y''|y|^a||_{\infty} \int_{x_0}^{x_n} y'(x)|y|^{-a}(x)dx \\
&\le |y'(x_n)|^2 + K||y''|y|^a||_{\infty}||y||_{\infty}^{1-a}.
\end{aligned}$$

Letting $n \to \infty$ completes the proof for $J = R^+$.

In the case $J = R$ the arguments are similar with the additional observation that x_1 and x_n can always be chosen so that y has no zeros in (x_0, x_1) and (x_0, x_n) (or (x_1, x_0), (x_n, x_0)). Let us establish this claim for one case; the other cases are similar. Suppose y' has a zero both to the right and to the left of x_0. If $y'(x_0) = 0$, (1.70) holds. If $y'(x_0) \ne 0$, choose the two nearest zeros of y', $x_1 < x_0 < \bar{x}_1$, one on each side of x_0. We claim that either (x_1, x_0) or $(x_0, \bar{x}_1)$ contains no zero of y. Suppose not, and $t_1 \in (x_1, x_0)$ $t_2 \in (x_0, x_1)$ are zeros of y. Rolle's theorem then yields a zero of y' in (t_1, t_2) contradicting the choice of x_1 and $\bar{x}_1$. With this observation we see that (1.73) does not occur in this case when x_1 and x_n are properly chosen and (1.72) becomes

$$\int_{x_0}^{x_1} y'(x)|y|^{-a}(x)dx \le (1-a)^{-1}||y||_{\infty}^{1-a}.$$

Theorem 1.8 follows from this and (1.71). $\quad\square$

Remark 1.4 Theorem 1.8 is not valid (i.e., inequality (1.68) does not hold for any positive K) when $a = 1$. To see this consider the initial value problem

$$y''y = -2, \quad y(0) = 1, \quad y'(0) = 0. \qquad (1.75)$$

The solution of (1.75) is given implicitly by

$$t = \int_y^1 (-\log y)^{-1/2} dy$$

on $[0, t_0]$ with $t_0 = \int_0^1 (-\log y)^{-1/2} dy$. Now extend y to $[-t_0, t_0]$ as an even function and then to the whole real line as a periodic function of period $2t_0$. In $[0, t_0]$, we have $y' < 0$, $y'^2 = (-\log y)$. Hence y is decreasing on $[0, t_0]$ and so $\|y\|_\infty = y(0) = 1$. On the other hand y' is not bounded since $\lim |y'(t)| = \infty$ as $t \to t_0$.

1.5   Notes and Problems

The material in Section 1 is all elementary and surely must be known, although we have not seen a statement of Theorem 1.1 in the literature.

Section 2. Theorem 1.2 and the three lemmas are well known — see [Kwong and Zettl] [1980a] [although Lemma 1.2 is stated more generally than one finds in the literature]. The proof is standard. Such inequalities are basic to the study of Sobolev imbedding theorems and are important in the theory of partial differential equations. See Adams [1975] and Friedman [1969]. Theorem 1.3 is also known [Ljubic 1964], but our proof is different than the one given by Ljubic. It is more elementary. We mention some interesting questions and problems.

1. Does inequality (1.21) extend to the case when the three norms are different? In other words, for what values of p, q, and r do we have

$$\|y^{(k)}\|_q < \epsilon \|y^{(n)}\|_r + K(\epsilon)\|y\|_p? \tag{1.76}$$

Although Lemma 1.2 holds for different norms, our proof of Lemma 1.3 uses the additivity of the integral. This is the obstacle one faces in trying to go from the small interval case to the "large" interval case in the proof of Lemma 1.2.

2. Let $\mu = \mu(k, n, p, J)$ denote the smallest constant in the inequality

$$\|y^{(k)}\|_p \leq \mu [\|y\|_p + \|y^{(n)}\|_p]. \tag{1.77}$$

What are the (exact) values of $\mu(k, n, p, J)$, $1 \leq k < n$, $1 \leq p \leq \infty$, and J any interval bounded or unbounded? The answer seems to be known for only a few cases. For $p = 2$ and $J = R^+$ these constants can be computed by the Ljubic-Kupcov algorithm. V. Q. Phong [1981] has developed an algorithm to compute these constants for $p = 2$ and J a bounded interval. He implemented it only for one case and found

$$\mu(1, 2, 2, [0, 1]) \approx 6.45. \tag{1.78}$$

3. When does there exist an extremal for (1.77)? Recall that an extremal is a nontrivial function y for which equality exists in (1.77). It is known that extremals exist for $n = 2$, $k = 1$, and all $p = q = r$, $1 < p \leq \infty$ when $J = R^+$ but not when $J = R$. This was shown by Kwong and Zettl [1980a, Theorem 6.1, p. 204] for a related inequality, i.e., (1.25), but this case follows from that one.

4. Given positive numbers u_0, u_k, u_n satisfying $u_k < \mu(k, n, p, J)(u_0 + u_n)$, does there exist a function $y \in W_{p,p}^n(J)$ such that

$$\|y\|_p = u_0, \quad \|y^{(k)}\|_p = u_k, \quad \|y^{(n)}\|_p = u_n?$$

It follows from Theorem 1.6 that the answer is yes when J is unbounded. For J bounded the answer in general is no.

In this case a more interesting question is:

5. Suppose $J = [a, b]$ is compact. Let u_0, u_n be positive numbers. Consider the set $S(u_0, u_n)$ of all $y \in W_{p,r}^n(J)$, $1 \leq p, r \leq \infty$, $n = 2, 3, \ldots$ satisfying

$$\|y\|_p = u_0 \text{ and } \|y^{(n)}\|_r = u_n. \tag{1.79}$$

For $1 \leq q \leq \infty$, k an integer with $1 \leq k < n$, what values can $\|y^{(k)}\|_q$ assume as y varies over $S(u_0, u_n)$?

Chui and Smith [1975] considered this question for the case $p = q = r = \infty$, $J = [0, 1]$ (any compact interval case can be reduced to this), $n = 2$, $k = 1$, $u_0 = 1$ (the general case for u_0 can be reduced to this by replacing y by cy, where c is an appropriate constant). Let $u_2 = u$, $\|\cdot\| = \|\cdot\|_\infty$.

Theorem *(Chui and Smith [1975]). If $\|y\| = 1$ and $\|y''\| = u$, then*

$$\|y'\| \leq \begin{cases} (u + 4)/2 & \text{if } 0 \leq u \leq 4 \\ 2\sqrt{u} & \text{if } 4 < u < \infty. \end{cases} \tag{1.80}$$

Furthermore, let u_1 be any number such that

$$\begin{aligned} 0 \leq u_1 \leq 2 & \quad \text{if } u = 0 \\ 0 < u_1 \leq (u + 4)/2 & \quad \text{if } 0 < u \leq 4 \\ 0 < u_1 \leq 2\sqrt{u} & \quad \text{if } u > 4. \end{aligned} \tag{1.81}$$

Then there exists a function y in S such that

$$\|y'\| = u_1.$$

Remark. Actually Chui and Smith did not prove the furthermore part of this theorem but they did show that the upper bounds in (1.80) are best possible by explicitly constructing an extremal function. This result was also considered by Sato and Sato [1983]. Also, M. Sato [1983] investigated the case $\|y'''\| = u$.

The proof we give now is different from that in Chui and Smith [1975]. We believe it extends to the case $1 \leq p < \infty$, to the extent of showing that the norm of the derivative can assume any nonzero value less than the least upper bound. But it might not be easy to determine the least upper bound of $\|y'\|_p$ in terms of p and u. In any case we do not pursue this matter further here for $p < \infty$.

Proof. To explain why the case $u = 0$ is an exception in (1.81) we note that if $u_1 = 0$ then $y'(t) = 0$ for all $t \in J$ implying that $y''(t) = 0$ for all $t \in J$ and so $u = 0$. Thus u_1 cannot be zero unless $u = 0$.

We first establish (1.81) by contradiction. Naturally there are two cases. Let $0 \leq u \leq 4$. If $\|y'\| > (u+4)/2$, then at some point $c \in [0,1]$, $|y'(c)| > (u+4)/2$. Without loss of generality we may assume that $y'(c) > (u+4)/2$. Thus for $t \in (c,1]$

$$y'(t) = y'(c) + \int_c^t y''(s)ds > \frac{u+4}{2} - u(t-c) \tag{1.82}$$

and hence

$$
\begin{aligned}
y(1) - y(c) &= \int_c^1 y'(t)dt > \int_c^1 \frac{u+4}{2} - u(t-c)dt \\
&= \frac{(u+4)(1-c)}{2} - \frac{u(1-c)^2}{2}.
\end{aligned}
\tag{1.83}
$$

For $t \in [0,c)$,

$$y'(t) = y'(c) - \int_t^c y''(s)ds > \frac{u+4}{2} - u(c-t) \tag{1.84}$$

and hence

$$y(c) - y(0) > \frac{(u+4)}{2}c - \frac{uc^2}{2}. \tag{1.85}$$

Adding (1.83) and (1.85) gives $y(1) - y(0) > 2 + u(c - c^2) \geq 2$. This contradicts the fact that $|y(1) - y(0)| \leq |y(1)| + |y(0)| \leq 2\|y\| = 2$.

Now suppose $u > 4$ and $y'(c) > 2\sqrt{u}$ for some $c \in [0,1]$. Let $[\alpha, \beta]$ be a subinterval of $[0,1]$ of length $2/\sqrt{u}$ that contains c. Since $2/\sqrt{u} < 1$, such an interval exists.

Estimating y' on $[c, \beta]$ as in (1.84) and then $y(\beta)$ as in (1.83) gives

$$y(\beta) - y(c) > 2\sqrt{u}(\beta - c) - \frac{u(\beta - c)^2}{2}.$$

The analogue of (185) is

$$y(c) - y(\alpha) > 2\sqrt{u}(c - \alpha) - \frac{u(c - \alpha)^2}{2}.$$

Hence

$$
\begin{aligned}
y(\beta) - y(\alpha) &> 2\sqrt{u}(\beta - \alpha) - \frac{u}{2}[(\beta - c)^2 + (c - \alpha)^2] \\
&= 4 - \frac{u}{2}[(\beta - c)^2 + (c - \alpha)^2].
\end{aligned}
$$

If we let $x = \beta - c$ and $y = c - \alpha$, then $x + y = 2/\sqrt{u}$. It is not hard to see that $x^2 + y^2$ attains its maximum when either $x = 0$ or $y = 0$. Thus $[(\beta - c)^2 + (c - \alpha)^2] \le \dfrac{4}{u}$. This gives the same contradiction $y(\beta) - y(\alpha) > 2$ as in the previous case.

As pointed out in Chui and Smith [1975], in each case, the maximum value of $\|y'\|$ allowed by (1.81) is attainable, e.g., by the following "extremal" functions

$$y_1(t) = \tfrac{1}{2}[ut^2 + (4 - u)t - 2] \qquad \text{if } 0 \le u \le 4$$

$$y_1(t) = \begin{cases} -1 & 0 \le t \le s \\ \tfrac{1}{2}u(t - s)^2 - 1 & t > s \end{cases} \qquad \text{if } u > 4 \text{ where } s = 1 - 2/\sqrt{u}.$$

To see that arbitrarily small values of $\|y'\|$ are also attainable, consider, for a fixed u, the function

$$y_2(t) = -1 + \frac{\epsilon^2}{u}(1 - \cos \frac{u}{\epsilon}t).$$

Then if ϵ is small enough, $\|y_2\| = 1$, $(|y_2(0)| = 1, |y_2(t)| \le 1)$, $\|y_2'\| = \epsilon$, and $\|y_2''\| = u$.

Consider convex combinations of y_1 and y_2:

$$y_\lambda(t) = \lambda y_1(t) + (1 - \lambda)y_2(t) \qquad 0 \le \lambda \le 1.$$

First notice that $\|y_\lambda\| \le \lambda\|y_1\| + (1 - \lambda)\|y_2\| = 1$. Similarly $\|y_\lambda''\| \le u$. It is not hard to see that in fact $\|y_\lambda\| = 1$ (the maximum of $|y_\lambda(t)|$ is attained at $t = 0$) and $\|y''\| = u$ (there exists some point t in a neighborhood of 1 at which both $y_1''(t)$ and $y_2''(t)$ are u). By continuity all values between ϵ and $\|y_1\|$ are attained by some y_λ.

For the general case $1 \le p < \infty$ the least upper bound of the values of $\|y'\|_p$ seems not to be known. However, it can be shown by similar but more technical arguments that $\|y'\|_p$ can assume any value between 0 and the least upper bound. $\square$

Section 3.

1. The basic question here is: Let (1.26) and (1.27) hold. What are the (exact) values of the smallest constant $K = K(n, k, p, q, r, J)$, $J = R$ or $J = R^+$, in the inequality (1.25)? This question has a long history going back at least to a paper of Landau [1913]

and has been worked on by many mathematicians including Hadamard [1914], Shilov [1937], Kolmogorov [1962], Hardy and Littlewood [1932], Schoenberg and Cavaretta [1970], Steckin [1965], Arestov [1967, 1972a, 1972b], Gabushin [1967], Ditzian [1977], Ljubic [1964], Hille [1970, 1972], Berdyshev [1971], and many others. Yet the answer is known only in a few special cases: mainly when $p = q = r = 1, 2$, or ∞. The best constants are summarized in the Appendix.

2. A question related to 1 is: When do extremals exist? In all but one case for which extremals are known in (1.25), the best value of K was explicitly known and the extremals explicitly exhibited. The exceptional case is found in [Kwong and Zettl 1980a] where it is shown that extremals exist for the case $n = 2$, $k = 1$, $1 < p = q = r < \infty$, $J = R^+$ even though the constants $K(2, 1, p, p, p, R^+)$ are not known, i.e., only an existence proof is given. It is also shown in [Kwong and Zettl 1980a] that there are no extremals for $K(2, 1, p, p, p, R)$.

3. Are the extremals for $K(2, 1, p, p, p, R^+)$ essentially unique, i.e., unique modulo $y(t) \rightarrow ay(bt)$? Are these extremals oscillatory for all p, $1 < p < \infty$? Hardy and Littlewood [1932] showed that the answer to both questions is yes when $p = 2$.

4. Does $K(n, k, p, p, p, R) = K(n, k, q, q, q, R)$ when $p^{-1} + q^{-1} = 1$? Ditzian [1975] showed that the answer is yes when $p = 1$, $q = \infty$.

5. It is shown in Kwong and Zettl [1980a] that the constants $K(2, 1, p, p, p, J)$ depend continuously on p, $1 \leq p \leq \infty$. Do the constants $K(n, k, p, q, r, J)$ depend continuously on p, q, r, $1 \leq p, r \leq \infty$? It would be very surprising if this were not true.

6. Is $K(2, 1, p, p, p, R^+)$ increasing for $2 \leq p \leq \infty$? decreasing for $1 \leq p \leq 2$? We conjecture that the answer is yes.

7. Berdyshev [1971] found that $K(2, 1, 1, 1, 1, R^+) = \sqrt{5/2}$. What are the values of $K(n, k, 1, 1, R^+)$ for $1 \leq k < n$, $n = 3, 4, \ldots$?

Problem 1. In the absence of knowledge of the exact value of $K(n, k, p, q, r, J)$, find "good" upper and lower bounds. For work along these lines see the paper by Franco, Kaper, Kwong, and Zettl [1983].

8. Inequalities of type (1.25) for functions of more than one variable have also been studied. See Konavalov [1978] and some of the references therein.

9. **Connections with approximation theory.** Steckin [1967] found a connection between the best constant in (1.25) and the problem of approximating the unbounded differentiation operator by bounded operators.

Let $Q = \{y \in W_{p,p}^n(K) : \|y^{(n)}\|_p \leq 1\}$, $1 \leq p \leq \infty$, and let B_N denote the set of bounded linear operators on $L^p(J)$ with bound $\leq N$. Define

$$E(N, J) = E(N) = \inf\{U(T) : T \in B_N\},$$

where

$$U(T) = \sup\{\|y^{(k)} - Tf\|_p : f \in Q\}, \qquad 1 \leq k < n.$$

Intuitively $U(T)$ measures how well a particular bounded operator T approximates the kth power of the differentiation operator d/dt over the set Q, and $E(N)$ measures the best approximation of d^k/dt^k by bounded operators of norm $\leq N$.

Theorem *(Steckin [1967]) Let $1 < r = p = q \leq \infty$, let n, k be integers with $1 \leq k < n$, and let α, β be given by (1.26), (1.27). Then for $J = R$ or R^+*

$$\alpha^\alpha \beta^\beta K(n, k, p, p, p, J) \leq E^\beta(1, J).$$

There is a vast Soviet literature on the connection between the best constant in (1.25) and the problem of approximating one class of operators by another. The interested reader is referred to the article by Arestov [1972b].

10. Nagy [1941] studied the case $n = 1$, $k = 0$ of (1.25) and found the exact values of the constants K for different p norms.

Theorem *(Nagy [1941]) Let $0 < p < \infty$, $1 \leq r < \infty$, $J = R$. If $y \in W_{p,r}^1(R)$, then for $q > p$ and $c = 1 + p(1 - r^{-1})$ we have*

$$\|y\|_\infty \leq (c/2)^{1/c} \|y\|_p^{p(r-1)/(rc)} \|y'\|_r^{1/c}$$
$$\|y\|_q \leq \left[c/2\, H\left(c/(q-p), 1 - 1/r\, \|y\|_p^{p(1+(q-p)(r-q)/rc)/q} \|y'\|_r^{(q-p)/cq}\right)\right],$$

where

$$H(u, v) = \frac{(u+v)^{-(u+v)}\Gamma(1 + u + v)}{u^{-u}\Gamma(u)v^{-v}\Gamma(v)}, \qquad H(u, 0) = H(0, v) = 1,$$

and the constants in both inequalities are sharp.

Chapter 2

The Norms of y, y', y''

2.1 Introduction

In this chapter, we discuss the relationship between the norms of a function and its first two derivatives. The reason for specializing to the second-order case is primarily because (i) it is of special interest and (ii) much more is known in this case.

Taking $n = 2$, $k = 1$ in inequality (1.25) of Section 1.3 we have

$$\|y'\|_q \leq K \, \|y\|_p^\alpha \, \|y''\|_r^\beta, \tag{2.1}$$

where $1 \leq p, q, r \leq \infty$,

$$\alpha = (1 - r^{-1} + q^{-1})/(2 - r^{-1} + p^{-1}), \qquad \beta = 1 - \alpha, \tag{2.2}$$

and

$$2q^{-1} \leq p^{-1} + r^{-1}. \tag{2.3}$$

Letting $p = q = r$ in (2.1) and squaring gives

$$\|y'\|_p^2 \leq K^2 \|y\|_p \, \|y''\|_p. \tag{2.4}$$

For brevity the smallest constant in (2.4) is denoted by $k(p, J) = K^2(p, J) = K^2(2, 1, p, p, p, J)$. The only known (exact) values of $K^2(p, J)$ are when $p = 1, 2, \infty$. These are discussed in the next sections.

2.2 The L^∞ Case

For $p = \infty$ the best constants in inequality (2.4) were found by Landau [1913] when $J = R^+$ and by Hadamard [1914] in the whole line case.

Theorem 2.1 *(Landau, Hadamard)*

(a) $k(\infty, R^+) = 4$

(b) $k(\infty, R) = 2$

(c) There exist extremals in both cases $J = R$ and $J = R^+$ and these are not essentially unique.

   Proof. It was shown in Theorem 1.8 of Chapter 1 that $k(\infty, R^+) \leq 4$ and $k(\infty, R) \leq 2$. Thus it suffices to construct nontrivial functions for which equality occurs in (2.4) with these values of k. The nonuniqueness of extremals will also be clear from this construction.

   It can be verified directly that the following functions are extremals for the whole line case and the half line case, respectively:

$$y_1(t) = \begin{cases} t(2-t) & 0 \leq t \leq 2 \\ (t-2)(t-4) & 2 \leq t \leq 4 \end{cases} \quad \text{periodic of period 4} \qquad (2.5)$$

and

$$y_2(t) = \begin{cases} -\frac{1}{4} + t - \frac{1}{2}t^2 & 0 \leq t \leq 1 \\ \frac{1}{4} & 1 \leq t \leq \infty. \end{cases} \qquad (2.6)$$

By redefining y_1 in (2.5) to be the constant 1 on $(1, \infty)$ and -1 on $(-\infty, -1]$ we get another extremal for the whole line case. By redefining y_2 in (2.6) on the interval $[2, \infty)$ we can construct other extremals for the half line case. This completes the proof of Theorem 2.1. □

   We remark that the proof that $K^2 \leq 4$ when $J = R^+$ and $K^2 \leq 2$ when $J = R$ given in Theorem 1.8 of Section 1.4 is different from Landau's and Hadamard's original proofs which are based on Taylor's Theorem.

2.3   The L^2 Case

Hardy and Littlewood [1932] found the best constant and all extremals in the half-line case when $p = 2$. The (easy) whole line case is discussed in the classic book by Hardy, Littlewood, and Polya [1934].

Theorem 2.2 *(Hardy-Littlewood)*

(a) $k(2, R^+) = 2$ *and* $y(t) = \exp(-t/2)\sin(\sqrt{3}t/2 - \pi/3)$, $t > 0$ *is an extremal. All other extremals are given by* $a\,y(bt)$, $a \in R$, $b \in R^+$.

(b) $k(2, R) = 1$ *and there are no extremals.*

Proof. Part (a) has several known proofs. The classical one found in Hardy, Littlewood, and Polya [1934] uses the Calculus of Variations. We give here one of the simplest proofs. Let $f, f' \in L^2(0, \infty)$. By Lemma 1.5 of Chapter 1, Section 3, $\lim_{t \to \infty} f(t) = 0$. Hence

$$
\begin{aligned}
\int_0^\infty f(t)f'(t)dt &= \lim_{T \to \infty} \int_0^T f(t)f'(t)dt \\
&= \lim_{T \to \infty} \frac{1}{2}[f^2(T) - f^2(0)] \\
&= -\frac{1}{2}f^2(0) \\
&\leq 0.
\end{aligned}
$$

Let $y, y', y'' \in L^2$. It is easy to verify the identity

$$
c^2|y'|^2 + (c^2y'' + cy' + y)^2 = c^4|y''|^2 + |y|^2 + 2c(cy' + y)'(cy' + y)
$$

for any real constant c.

Integrating over $[0, \infty)$ gives

$$
c^2\|y'\|_2^2 + \|c^2y'' + cy' + y\|_2^2 = c^4\|y''\|_2^2 + \|y\|_2^2 + 2c\int_0^\infty (cy' + y)'(cy' + y)dt. \qquad (2.7)
$$

As seen above, the last integral is nonpositive. Thus we have the inequality

$$
c^2\|y'\|_2^2 \leq c^4\|y''\|_2^2 + \|y\|_2^2.
$$

Choosing $c^2 = \|y\|_2/\|y''\|_2$ gives

$$
\|y'\|_2^2 \leq 2\|y\|_2\|y''\|_2.
$$

Thus $k(2, R^+) \leq 2$. Equality holds if the terms left out from (2.7) are zero, namely if

$$
c^2y'' + cy' + y = 0
$$

and

$$
cy'(0) + y(0) = 0.
$$

Solving this initial value problem gives the extremals as stated in the theorem. Since equality can hold, $k(2, R^+) = 2$.

To prove part (b), let $y, y', y'' \in L^2(-\infty, \infty)$. Integration by parts gives

$$
\int_{-\infty}^\infty (y'(t))^2 dt = -\int_{-\infty}^\infty y(t)y''(t)dt.
$$

Schwarz's inequality gives

$$
\|y'\|_2^2 \leq \|y\|_2\|y''\|_2.
$$

This shows that $k(2,R) \le 1$. To show the opposite inequality, we construct the following sequence of test functions

$$y_n(t) = \begin{cases} \sin t & t \in [-n\pi, n\pi] \\ \text{smoothing part} & t \in [-n\pi - 1, -n\pi] \cup [n\pi, n\pi + 1] \\ 0 & \text{otherwise.} \end{cases}$$

The smoothing part is added to make y_n a C^∞ function, and is the same (after translation) for all n. It is easy to see that

$$\lim_{n \to \infty} \frac{\|y_n'\|_2^2}{\|y_n\|_2 \|y_n''\|_2} = 1$$

from which we deduce that $k(2,R) \ge 1$. The nonexistence of extremals is a special case of the general theorems in Section 2.8 below. $\square$

2.4   Equivalent Bounded Interval Problems for R

We remarked before that an inequality of type (2.1) cannot hold on a bounded interval J unless the class of admissible functions is restricted by imposing further conditions. For a further discussion of this point see Section 1.3. Here we investigate the type of end point conditions for which (2.4) holds with a finite constant k and for which this constant $k = k(p,R)$, i.e., is the same as the whole line constant.

First we make some general remarks about the effects of "scaling" and introduce some notation. In place of $W_{p,p}^n(J)$ we write $W_p^n(J)$. For $y \in W_p^n(J)$ with $y \neq 0$, $y'' \neq 0$ let

$$G(y) = \|y'\|_p^2 / (\|y\|_p \|y''\|_p). \tag{2.8}$$

For M a subset of $W_p^n(J)$ let

$$k(p,M) = \sup\{G(y) : y \in M, y \neq 0, y'' \neq 0\}.$$

In general, for M a nonzero subset of $W_p^n(J)$, $k(p,M)$ is a positive number or $+\infty$. If $k(p,M)$ is positive then $k = k(p,M)$ is the smallest constant k in (2.4) for all $y \in M$. Also $k(p, L^p(J)) = k(p,J)$, $J = R$ or R^+.

The quotient $G(y)$ of (2.8) is invariant under each of the following "scaling" transformations:

(i) horizontal scaling: replacing t by at for any constant $a \neq 0$.

(ii) vertical scaling: replacing y by a scalar multiple by, $b \neq 0$.

(iii) translation along the real axis: replacing t by $t + h$ for any constant h.

In particular, the quotient $G(y)$ and hence the constant $k(p, M)$ is independent of the compact interval $J = [a, b]$. That is, if y is defined on some compact interval $[a, b]$, we may, by horizontal scaling and translation, assume that y is defined on $[0, 1]$ (or any other compact interval), i.e., we may replace y by a function z defined on $[0, 1]$ such that $G(y) = G(z)$. This observation will be used often and repeatedly below, sometimes without explicit mention.

We consider the following sets of functions for $1 \leq p \leq \infty$:

$$M_1 = \{y \in W_p^2([0, 1]) : y(0) = y(1) = 0\}$$
$$M_2 = \{y \in M_1 : y(t) > 0, 0 < t < 1\}$$
$$M_3 = \{y \in W_p^2([0, 1]) : y(0) = y'(1) = 0, y(t) > 0, 0 < t < 1\}$$
$$M_4 = \{y \in M_3 : y'(t) \geq 0, 0 < t < 1\}$$
$$M_5 = \{y \in W_p^2([0, 1]) : y'(0) = 0 = y'(1)\}$$
$$M_6 = \{y \in W_p^2(R^+) : y(0) = 0\}$$
$$M_7 = \{y \in W_p^2(R^+) : y'(0) = 0\}$$
$$M_8 = \{y \in W_p^2(R^+) : y'(0)y(0) \geq 0\}$$
$$M_9 = \{y \in W_p([0, 1]) : y'(1) = 0, y'(t) \geq 0, y(t) \geq 0, 0 < t < 1\}$$

Theorem 2.3 *For any p, $1 \leq p < \infty$,*

$$k(p, M_i) = k(p, R), \quad i = 1, 2, \ldots, 9. \tag{2.9}$$

Furthermore, the interval $[0, 1]$ in the definition of M_i, $i = 1, 2, \ldots, 9$ can be replaced by any compact interval $[a, b]$ and the end point conditions transferred accordingly.

Proof. The "furthermore" statement follows from the observations made in the paragraph preceding the definition of the M_i.

The proof of (2.9) is based on two lemmas. The first one is a special case of Lemma 1.6 of section 3 in Chapter 1, and is stated here only for the convenience of the reader. $\square$

Lemma 2.1 *Let $y \in W_p^2(R)$, $1 \leq p < \infty$. For any $\epsilon > 0$ there exists a $z \in C_0^\infty(R)$ such that*

$$\|y - z\|_p < \epsilon, \quad \|y' - z'\|_p < \epsilon, \quad \|y'' - z''\|_p < \epsilon. \tag{2.10}$$

Proof. See Lemma 1.6 of Section 3 in Chapter 1. $\square$

The next lemma plays a fundamental role in this section and several subsequent ones. It involves the W_p^2 spaces over unions of intervals I_i. By

$$y \in W_p^2(U_{i=1}^n I_i)$$

we mean that the restriction of y to I_i is in $W_p^2(I_i)$ for each $i = 1, \ldots, n$.

Lemma 2.2 *Let* I_i, $i = 1, \ldots, n \geq 2$ *be a finite number of intervals, one or two of which may be unbounded, having at most end points in common. Let*

$$h \in W_p^2(\bigcup_{i=1}^{n} I_i)$$

and denote the restriction of h *to* I_i *by* h_i. *Then there exists a* $j \in \{1, \ldots, n\}$ *such that*

$$G(h_j) > G(h) \tag{2.11}$$

unless the 3 n-tuples

$$(\|h_i\|_p)_{i=1}^n, \quad (\|h_i'\|_p)_{i=1}^n, \quad (\|h_i''\|_p)_{i=1}^n$$

are proportional, in which case

$$G(h_i) = g(h), \ \text{for all} \ i = 1, \ldots, n. \tag{2.12}$$

Proof. It suffices to establish the case $n = 2$. Let

$$A = \|h_1\|_p^p, \quad B = \|h_1''\|_p^p, \quad C = \|h_2\|_p^p, \quad D = \|h_2''\|_p^p, \quad a = G(h).$$

Then

$$\|h\|_p^p = A + C, \quad \|h''\|_p^p = B + D$$

and

$$\|h'\|_p^{2p} = a^p \|h\|_p^p \|h''\|_p^p = a^p(A + C)(B + D).$$

By the Schwarz inequality in two-space

$$AB + 2(ABCD)^{1/2} + CD \leq (A + C)(B + D)$$

with equality if and only if $(A^{1/2}, C^{1/2})$ and $(B^{1/2}, D^{1/2})$ are proportional, i.e.,

$$A = \alpha B, \quad C = \alpha D$$

for some constant α.

Suppose $G(h_i) \leq a$ for $i = 1, 2$ and the pairs $(A, C), (B, D)$ are not proportional. Then

$$
\begin{aligned}
\|h'\|_p^{2p} &= (\|h_1'\|_p^p + \|h_2'\|_p^p)^2 \leq a^p[(AB)^{1/2} + (CD)^{1/2}]^2 \\
&= a^p[AB + 2(ABCD)^{1/2} + CD] < a^p(A + C)(B + D) \\
&= a^p \|h\|_p^p \|h''\|_p^p = \|h'\|_p^{2p}.
\end{aligned}
$$

This contradiction shows that if the pairs (A, C) and (B, D) are not proportional, then $G(h_i) < G(h)$ for at least one i and completes the proof of Lemma 2.2. $\square$

Proof. **of Theorem 2.3** To establish (2.9) for $i = 1$, let $\epsilon > 0$. There exists an f in $W_p^2(R)$ such that

$$k(p, R) - \epsilon < G(f).$$

By Lemma 2.1 there exists a $g \in C_0^\infty(R)$ such that

$$G(f) < G(g) + \epsilon.$$

Hence

$$k(p, M_1^*) \geq G(g) > k(p, R) - 2\epsilon,$$

where $M_1^* = \{y \in W_p^2([a, b]) : y(a) = y(b) = 0\}$ and the compact interval $[a, b]$ is chosen to contain the support of g. But M_1^* can be replaced by M_1 by the remarks about scaling made earlier. Letting $\epsilon \to 0$ we conclude that

$$k(p, M_1) \geq k(p, R).$$

To prove the reverse inequality, let $\epsilon > 0$ and choose $g \in M_1$ such that

$$k^p(p, M_1) - \epsilon < G^p(g).$$

Define h on $[-1, 0]$ such that h is zero in a right neighborhood of -1 and h together with g form a function in $W_p^2([-1, 1])$. For any positive integer $n \geq 2$ define a function f_n in $W_p^2(R)$ as follows: $f_n(t) = 0$ for $t \leq -1$, $f_n(t) = h(t)$, $-1 \leq t \leq 0$, $f_n(t) = g(t)$, $0 \leq t \leq 1$, $f_n(t) = -g(2 - t)$, $1 \leq t \leq 2$, f_n in $[2j - 2, 2j]$ is a copy of f_n in $[0, 2]$ for $j = 1, \ldots, n$ and f_n in $[2n, 2n + 1]$ is a copy of h and finally $f_n(t) = 0$ for $t \geq 2n + 1$. Geometrically f_n is n copies of an odd extension of g smoothed out at both ends so as to be a smooth function on R.

¿From this construction it follows that

$$\|f_i^{(i)}\|_p^p = 2n\|g^{(i)}\|_p^p + 2A_i, \qquad i = 0, 1, 2,$$

where $A_i = \|h^{(i)}\|_p^p$.

Thus

$$\begin{aligned}(G(f_n))^p &= (\|g'\|_p^p + A_1/n)^2/((\|g\|_p^p + A_0/n)(\|g''\|_p^p + A_2/n)) \\ (G(g))^p &> k^p(p, M_1) - \epsilon \quad \text{as } n \to \infty.\end{aligned}$$

Hence

$$k^p(p, R) \geq (G(f_n))^p > k^p(p, M_1) - 2\epsilon.$$

Letting $\epsilon \to 0$ and taking p^{th} roots we conclude that

$$k(p, R) \geq k(p, M_1)$$

and hence

$$k(p, R) = k(p, M_1)$$

completing the proof of Theorem 2.3 for $i = 1$. Let $k_i = k(p, M_i)$, $i = 1, \ldots, 9$. To establish the case $i = 2$ it suffices to show that $k_2 = k_1$. Clearly $k_2 \leq k_1$. To prove that $k_1 \leq k_2$, let $\epsilon > 0$ and choose $h \in M_1$ such that

$$G(h) > k_1 - \epsilon.$$

First we consider the case when h has a finite number of zeros, say $0 \leq t_0 < t_1 < \ldots < t_{m+1} = 1$. Let h_j be the restriction of h to $[t_{j-1}, t_j]$. By Lemma 2.2, $G(h_j) > k_1 - \epsilon$ for some j. Since the quotients $G(y)$ are independent of the interval we can conclude that

$$k_2 \geq G(|h_j|) > k_1 - \epsilon.$$

Next we consider the case when h has an infinite number of zeros in $[0, 1]$. Then the open set $\{t \in [0, 1] : h(t) \neq 0\} = \cup_{k=1}^{\infty} I_k$, where the I_k's are disjoint open (in the relative topology of $[0, 1]$) intervals in $[0, 1]$. ¿From the additivity of the Lebesgue integral we see that

$$\lim_{m \to \infty} \sum_{k=1}^{m} \int_{I_k} |h^{(j)}|^p = \|h^{(j)}\|_p^p, \quad j = 0, 1, 2.$$

Hence for m sufficiently large

$$\left(\sum_{k=1}^{m} \int_{I_k} |h'|^p \right)^2 \bigg/ \left(\sum_{k=1}^{m} \int_{I_k} |h|^p \right) \left(\sum_{k=1}^{m} \int_{I_k} |h''|^p \right) > k_1 - 2\epsilon.$$

Now defining h_j to be the restriction of h to I_j and applying Lemma 2.2 as above, we conclude that $k_2 > k_1 - 2\epsilon$.

Finally, if h has no zero on $(0,1)$ and is negative, we may replace h with $-h$. This completes the proof for $k_2 = k_1$. $\square$

Now to show that $k_3 = k_2$ take $I = [0, 1]$. Let $\epsilon > 0$ and choose $h \in W_p^2(I)$ such that $h \neq 0$, $h(0) = 0 = h(1)$, $h(t) \geq 0$, and $G(h) > k_2 - \epsilon$. Let $t_1 \in I$ so that $h'(t_1) = 0$. Then $0 < t_1 < 1$. Letting h_1 and h_2 denote the restrictions of h to $[0, t_1]$ and $[t_1, 1]$, respectively, and using Lemma 2.2 as above, we obtain $k_3 \geq k_2$. On the other hand, for $\epsilon > 0$ choose $g \in W_p^2(I)$ such that $g \neq 0$, $g(0) = 0 = g'(1)$, $g(t) > 0$, $0 < t < 1$, and $G(g) > k_3 - \epsilon$. Let $f = g$ on $[0,1]$ and define $f(t) = g(2 - t)$ for $1 \leq t \leq 2$. Then $G(g) = G(f)$. Finally, let $h(t) = f(2t)$. Then $G(g) = G(f) + G(h)$, $h(0) = 0 = h(1)$, $h(t) > 0$, $0 < t < 1$. Hence $k_2 \geq G(h) = G(g) > k_3 - \epsilon$. Consequently, $k_2 \geq k_3$ and $k_2 = k_3$.

Clearly $k_4 \leq k_3$. To prove the reverse inequality, let $\epsilon > 0$ and choose $h \in W_p^2(I)$ for $I = [0, 1]$ satisfying $h(0) = 0 = h'(1)$, $h(t) > 0$, $0 < t < 1$ so that $G(h) > k_4 - \epsilon$.

First we consider the case when h' has a finite number of zeros in $[0,1]$, say at the points $0 \leq t_1 < t_2 < \ldots < t_{m+1} = 1$. Let h_j be the restriction of h to $[t_{j-1}, t_j]$. Then, by Lemma 2.2, $G(h_j) > k_4 - \epsilon$ for some j. By rescaling (again using the fact that the quotients $G(y)$ are independent of the interval) we may assume that h_j is a function g defined on $[0,1]$ and satisfying $g'(1) = 0$ and $g'(t) > 0$ for $0 < t < 1$. But $g(0)$ need not be zero. Since $g'(t) > 0$, $0 < t < 1$ we have that $g(t) > g(0)$ for $t \in (0,1)$. Consider $f = g - g(0)$ and observe that $G(f) > G(g)$ since $\|f\|_p < \|g\|_p$ while $\|f'\| = \|g'\|$ and $\|f''\| = \|g''\|$. So $G(f) > G(g) = G(h_j) > k_4 - \epsilon$ and consequently $k_3 \geq k_4$ in this case. The case when h' has an infinite number of zeros can be reduced to the case with a finite number of zeros in a manner similar to the proof of $k_1 = k_2$ above.

The set M_9 differs from M_4 in that the functions in the former are not required to vanish at zero. Next we want to show that $k_9 = k_4$. Clearly $k_4 \leq k_9$ since $M_4 \subseteq M_9$. To prove the reverse inequality let $f \in M_9$. Then f is nondecreasing and $y = f - f(0)$ is in M_4. Since $\|y\|_p \leq \|f\|_p$ and $y^{(i)} = f^{(i)}$, $i = 1, 2$, it follows that

$$G(f) \leq G(y) \leq k_4.$$

Since for each $f \in M_9$ there is a $y \in M_4$ such that $G(f) \leq G(y)$ it follows that $k_9 \leq k_4$.

So far we have established the cases $i = 1, 2, 3, 4$ and 9. The remaining cases can be established similarly.

2.5 Equivalent Bounded Interval Problems for R^+

To get bounded interval characterizations of the constants $k(p, R^+)$, $1 \leq p \leq \infty$ we define

$$
\begin{aligned}
N_1 &= \{y \in W_p^2([0,1]) : y(1) = 0 = y'(1)\} \\
N_2 &= \{y \in W_p^2([0,1]) : \text{ either } y(0) = 0 = y(1) \text{ or} \\
&\quad y'(0)/y(0) = y'(1)/y(1) \text{ and } y' \text{ vanishes exactly once in} (0,1)\}.
\end{aligned}
$$

Theorem 2.4 *For any p, $1 \leq p \leq \infty$,*

$$k(p, R^+) = k(p, N_i), \quad i = 1, 2. \tag{2.13}$$

The constant $k(p, N_i)$ in (2.13) is the smallest constant k for which inequality (2.4) holds for all functions y in N_i.

Proof. The proof of the case $i = 1$ is similar to the proof of the corresponding case in Theorem 2.3 of Section 4 and so we only sketch it here. Let $f \in N_1$. Extend f to a function

$y \in W_p^2(R^+)$ by setting $y(t) = 0$ for $t \geq 1$. Then $G(f) = G(y)$. By taking the supremum we see that $k(p, R^+) \geq k(p, N_1)$. On the other hand, for any $\epsilon > 0$ there exists a $y \in W_p^2(R^+)$ such that $k(p, R^+) - \epsilon < G(y) \leq k(p, R^+)$. By Lemma 2.1 there exists a z in $C_0^\infty(R^+)$ such that $\|y - z\|$ and $\|y'' - z''\|$ and $\|y' - z'\|$ are so small that $G(z) \geq G(y) - \epsilon > k(p, R^+) - 2\epsilon$. Considering the restriction of z to its support and then scaling it appropriately we get a function $w \in N_1$ such that $G(w) = G(z) \geq k(p, R^+) - 2\epsilon$. Taking the supremum and then letting $\epsilon \to 0$ we find $k(p, N_1) \geq k(p, R^+)$.

To prove that $k(p, N_2) \leq k(p, R^+)$, choose $\epsilon > 0$ and take $f \in N_2$ such that $G(f) \geq k(p, N_2) - \epsilon$. If $f(0) = 0 = f(1)$, then $f \in M_1$ and so $G(f) \leq k(p, M_1) = k(p, R) \leq k(p, R^+)$. Now suppose $f(0) \neq 0$ and $f(1) \neq 0$. Then one of the following three cases occurs: (i) $f(0) > f(1)$, (ii) $f(0) < f(1)$, (iii) $f(0) = f(1)$. In the first case we construct a function $y \in W_p^2(R^+)$ by setting $y(t) = \alpha^n f(t - n)$, $t \in [n, n+1]$, where $\alpha = f(1)/f(0)$ and n is a positive integer. It follows that $G(y) = G(f)$. Thus $k(p, R^+) \geq G(y) \geq k(p, N_2) - \epsilon$. Letting $\epsilon \to 0$ completes case (i). The change of variable $x = 1 - t$ reduces case (ii) to case (i). In the third case, let $g(t) = f(t)(1 + ah(t))$, where h is a C^∞ function on $[0,1]$ which is 1 in a neighborhood of 0 and 0 is a neighborhood of 1. Then $g \to f$ as $a \to 0$ in the Sobolev norm of $W_p^2[0, 1]$. Hence, for a sufficiently small ϵ, $G(g) \geq G(f) - \epsilon \geq k(p, N_2) - 2\epsilon$. Now g is in case (i).

To establish the reverse inequality it suffices to show that $k(p, N_2) \geq k(p, N_1)$. For this it is enough to prove that for any $\epsilon > 0$ and $f \in N_1$ there exists a $g \in N_2$ such that $G(g) > G(f) - \epsilon$. First observe that if $f(0) = 0$ we can always approximate it by another function f_1 in N_1 such that $f_1(0) \neq 0$. Hence we may assume that $f(0) \neq 0$.

We consider several cases. If f has no zero in $(0,1)$ we may assume that $f(t) > 0$ for $t \in [0, 1)$, otherwise we replace f by $-f$. Let $h(t) = f(2t)$ on $[0,1/2]$. Let ϕ be a positive C^∞ function on $[1/2,1)$ such that $\phi(1) = \phi'(1) = 0$ and ϕ has the value 1 in a neighborhood of 1/2. For $\lambda > 0$ define g_λ on $[0,1]$ to be $h - \lambda$ on $[0,1/2]$ and $-\lambda\phi$ on $[1/2,1]$. The function g_λ has one zero, say x_λ, in $[0,1/2]$. Cleary x_λ is close to 1/2 for small λ. Also the Sobolev norm of $g_{\lambda,a}$, the restriction of g_λ to $[0, a]$ for any a in $[x_\lambda, 1]$ is close to that of h, uniformly in a. It follows that for any $\epsilon > 0$,

$$G(g_{\lambda,a}) \geq G(h) - \epsilon \tag{2.14}$$

for λ sufficiently small and for all a in $[x_\lambda, 1]$. Now we claim there is a point a in $[x_\lambda, 1]$ such that $g_\lambda'(a)/g_\lambda(a) = g'(0)/g_\lambda(0)$. To see this consider the function $F(t) = \log(-g_\lambda(t))$ on $(x_\lambda, 1)$. Then $F(t) \to -\infty$ as $t \to 1$ and as $t \to x_\lambda$. It follows from the mean value theorem that there are points near x_λ at which $F'(t)$ is as large as we like, in particular greater than $g_\lambda'(0)/g_\lambda(0)$; and, similarly, that there are points near 1 at which F' is less than $g_\lambda'(0)/g_\lambda(0)$. ¿From the continuity of F' we can then conclude that there is a point a in $(x_\lambda, 1]$

such that $F'(a) = g'_\lambda(a)/g_\lambda(a) = g'_\lambda(0)/g_\lambda(0)$. After scaling, the function $g_{\lambda,a}$ is in N_2 and $G(g) = G(g_{\lambda,a}) \geq G(h) - \epsilon = G(f) - \epsilon$.

Next we consider the case in which f has a finite number of zeros in $[0,1]$. Between any two such zeros, we can always find by the continuity argument used in the paragraph above a point a_i at which $f'(a_i)/f(a_i) = f'(0)/f(0)$. Such points divide $[0,1]$ into subintervals $[0,a_1], [a_1,a_2], \ldots, [a_n, 1]$. Let $f_1, \ldots, f_{n+1}$ be the restrictions of f on these intervals, respectively. Then by Lemma 2.2, one of these functions after being scaled gives a function g such that $G(g) \geq G(f)$. If g comes from $f_1, \ldots, f_n$ then $g \in N_2$ and the theorem is proved. If g comes from f_{n+1}, we are back to the case considered above.

We turn now to the remaining case in which f has an infinite number of zeros in $[0,1)$. Let a be the infimum of the set of zeros of f in $[0,1)$. By continuity $f'(a) = f(a) = 0$. By assumption $a > 0$. We apply Lemma 2.2 to f_1 and f_2, the restrictions of f to $(0,a)$ and $(a,1)$, respectively. If $G(f_2) \geq G(f)$, then, after scaling, f_2 yields a function g that vanishes at both end points. If $G(f_1) \geq G(f)$, then after scaling we get a function h in N_1 which has no accumulation point of zeros in $[0,1]$ other than perhaps 1. If 1 is not an accumulation point of zeros of h, we are back to Case 2 above. So suppose h has an infinite number of zeros in $[0,1]$ having 1 as the only accumulation point. Name the zeros in increasing order $x_1 < x_2 < \ldots < 1$, with $\lim_{n \to \infty} x_n = 1$. ¿From a property of integrals we see that given $\epsilon > 0$ there is a number X near 1 such that the Sobolev norm of h_a, the restriction of h to $[0,a]$, $a \in [X, 1]$ is so close to that of h that $G(h_a) \geq G(h) - \epsilon$. Let x_N be the first zero such that $x_N \geq X$. by the continuity argument employed in Case 1 above we can establish the existence of points $a_1 \in (x_i, x_{i+1})$, $i = 1, 2, \ldots, N$, such that $h'(a_i)/h(a_i) = h'(0)/h(0)$. After scaling, the restriction of h on each $[a_i, a_{i+1}]$, $i = 0, 1, \ldots, N$ (let $a_0 = 0$) yields a function in N_2. By the fundamental Lemma 2.2, one of these has a G value greater than or equal to $G(h_a)$ which is greater than or equal to $G(g) - \epsilon \geq G(f) - \epsilon$. This completes the proof of Theorem 2.4. $\square$

2.6   The L^1 Case

Berdyshev [1971] found the best constants in (2.4) when $p = 1$. The best constant for $k(1, R)$ was also found, independently, by Ditzian [1975].

Theorem 2.5 *(Berdyshev). We have*

  (a) $k(1, R^+) = 5/2$

(b) $k(1, R) = 2$

(c) *There is no extremal in either case.*

Proof. The proof we give here is due to Kwong and Zettl [1980a]. For $p = 1$ the problem M_4 does not have an extremal. The existence proof of Kwong and Zettl [1980a] for extremals in L^p, which we will discuss in Section 8 below, depends on the weak compactness of the unit ball in L^p and thus fails when $p = 1$. This failure can be surmounted by working in a larger space than W_1^2. Let $V(J) = \{f : f \text{ is the antiderivative of a function } g \text{ of bounded variation on } J \text{ and } f, g \in L(J)\}$, $J = [a, b]$, $-\infty \leq a < b \leq \infty$. The function G defined by (2.8) is now replaced by

$$G(f) = \|g\|_1^2 / (\|f\|_1 Vg), \tag{2.15}$$

where $\|f\|_1 = \int_J |f|$ and Vg is the total variation of g.

Note that the problem

$$M_{4,1} = \{f \in V([0,1]) : f(0) = 0, g(1) = 0,$$
$$f > 0 \text{ in } (0,1) \text{ and } g \geq 0 \text{ in } (0,1)\}$$

has an extremal in the extended sense based on G above. In fact such an extremal is given by

$$f(t) = t \text{ with } g(t) = \begin{cases} 1, & t \in [0, 1) \\ 0, & t = 1. \end{cases} \tag{2.16}$$

By arguments similar to those of Section 4 above, we may conclude that $k(1, R) = k(1, M_{4,1}) = 2$.

Similar remarks apply to problem N_2. Define $N_{2,1} = \{f \in V([0,1]): \text{ either } f(0) = f(1) = 0 \text{ or } g(0)/f(0) = g(1)/f(1), \text{ and } f \text{ vanishes exactly once in } (0,1)\}$.

Now we observe that

$$f(t) = \begin{cases} 1 - 2t & \text{in } [0, 3/4] \\ 2t/3 - 1 & \text{in } (3/4, 1] \end{cases} \tag{2.17}$$

with

$$g(t) = \begin{cases} -2 & \text{in } [0, 3/4] \\ 2/3 & \text{in } (3/4, 1] \end{cases} \tag{2.18}$$

is an extremal of problem $N_{2,1}$ in the extended sense based on (2.15). This gives, again by arguments similar to those of Section 4 above, that $k(1, R^+) = k(1, N_{2,1}) = 5/2$. □

For the half line problem R^+ an extremal (in the extended sense) in $V(R^+)$ can be obtained by extending the functions (2.17), (2.18) to the half line — see Berdyshev [1971].

No extremal exists for the whole line case — see Kwong and Zettl [1980a].

2.7   Upper and Lower Bounds for $k(p,R)$ and $k(p,R^+)$

Since the exact values of these constants are known only for $p = 1, 2, \infty$, we consider the problem of obtaining "good" upper and lower bounds. To do this it is useful to use some of the compact interval characterizations of these constants developed in Sections 4 and 5.

A lower bound can be obtained simply by evaluating $G(y)$ for a particular y from the appropriate set of functions.

Theorem 2.6 *For any* p, $1 \leq p \leq \infty$, *we have*

$$1 \leq k(p,R). \tag{2.19}$$

Proof. Consider $y(t) = \sin(\pi t/2)$, $t \in [0,2]$. Then $y \in M_1$ when the interval $[0,1]$ is replaced by $[0,2]$. A calculation shows that $G(y) = 1$ and (2.19) follows from Theorem 2.3. □

Theorem 2.7 *For any* p, $1 \leq p < \infty$, *we have*

$$k^p(p,R) \geq 2^{p+1}\Gamma(p+1+1/2)/[\pi^{1/2}\Gamma(p+1)\Gamma(p+2)] = (L(p))^p, \tag{2.20}$$

where Γ *denotes the gamma function.*

Proof. Consider

$$y(t) = t(1-t), \quad t \in [0,2].$$

Then y is in M_2 on the interval $[0,2]$. A calculation yields

$$\int_0^2 y^p = 2^{2p+1}\Gamma^2(p+1)/\Gamma(2p+2) = \sqrt{\pi}\Gamma(p+1)/\Gamma(p+1+1/2)$$
$$\int_0^2 |y'|^p = 2^{p+1}/(p+1)$$
$$\int_0^2 |y''|^p = 2^{p+1}.$$

Thus $G(y)$ is the right-hand side of (2.20) and the conclusion follows from Theorem 2.3. □

We will show in Section 9 that $k(p,R)$ is a continuous function of p. Since $k(2,R) = 1$, $k(\infty, R) = 2$, and 2 is the limit as $p \to \infty$ of the right-hand side of (2.20), it follows that (2.20) is a "good" lower bound when p is large and 1 is a good lower bound when p is near 2.

The next result improves on (2.19). The idea is to use a linear combination of the two functions employed in the proofs of Theorems 2.6 and 2.7 to improve the lower bound. Although this does not yield an explicit expression in terms of p it does show that strict inequality holds in (2.19) when $p \neq 2$.

Theorem 2.8 *If $1 \le p \le \infty$ and $p \ne 2$ then*

$$1 < k(p, R). \tag{2.21}$$

Proof. Consider the class of functions

$$f_x(t) = \sin(\pi t/2) + xt(2 - t), \qquad t \in [0, 1],$$

where $x \ge 0$ is a parameter. Note that $f_x \in M_4$ for $x \ge 0$. A calculation shows that f_0 is an extremal for $k(2, M_4)$. To establish (2.21) it will suffice to show that f_0 is not an extremal for $k(p, M_4)$ when $p \ne 2$. To do this let $A(x) = \|f_x\|_p^p$, $B(x) = \|f_x'\|_p^p$, $C(x) = \|f_x''\|_p^p$. We consider the derivatives with respect to x, d/dx, of A, B, C with the understanding that d/dx at $x = 0$ denotes the derivative from the right. Then at $x = 0$ we have

$$\begin{aligned}
\frac{dA}{dx} &= \int_0^1 pt(2 - t) \sin^{p-1}(\pi t/2) dt \\
\frac{dB}{dx} &= \int_0^1 2p(\pi/2)^{p-1}(1 - t) \cos^{p-1}(\pi t/2) dt \\
&= \int_0^1 2p(\pi/2)^{p-1}t \; \sin^{p-1}(\pi t/2) dt \\
\frac{dC}{dx} &= \int_0^1 2p(\pi/2)^{2p-2} \sin^{p-1}(\pi t/2) dt.
\end{aligned}$$

Next we analyze $\frac{d}{dx}G_p^p(f_x)$ at $x = 0$. The numerator is given by

$$ABC\frac{dB}{dx} - B^2(A\frac{dC}{dx} + C\frac{dA}{dx}) = (\pi/2)^{4p-2}A^3 \int_0^1 F(t)\sin^{p-1}(\pi t/2) dt, \tag{2.22}$$

where $F(t) = \pi^2 t^2/4 + (2\pi - \pi^2/2)t - 2$. Note that F is an increasing function of t in [0,1]. Since f_0 is an extremal for $k(2, M_4)$ we have $\frac{d}{dx}G_2^2(f_x) = 0$ at $x = 0$. This implies from (2.22) that $\int_0^1 F(t)\sin(\pi t/2) dt = 0$.

From this and the fact that $\sin t$ and $F(t)$ are both increasing on [0,1] it follows that

$$\int_0^1 F(t)\sin^{p-1}(\pi t/2) dt$$

is positive when $p > 2$ and negative when $p < 2$. Hence (2.22) is not zero when $p \ne 2$, i.e., $\frac{d}{dx}G_p^p(f_x) \ne 0$ at $x = 0$ when $p \ne 2$. Therefore, f_0 is not an extremal for $k(p, M_4)$ when $p \ne 2$ and inequality (2.21) holds. $\square$

Theorem 2.9 *For any p, $1 \le p \le \infty$, we have*

$$k(p, R) \le k(p, R^+). \tag{2.23}$$

Proof. We show that given any $h \in W_p^2(R)$ there exists a $g \in W_p^2(R^+)$ such that $G(h) \leq G(g)$ and (2.23) follows by taking the supremum. Let $h \in W_p^2(R)$ and denote by h_1, h_2 the restrictions of h to $(-\infty, 0)$ and $(0, \infty)$, respectively. Then $z(t) = h_1(-t)$ and h_2 are in $W_p^2(R^+)$. By Lemma 2 of Section 2 either $G(h_1) = G(z) \geq G(h)$ or $G(h_2) \geq G(h)$. □

In Section 8 we will see that strict inequality actually holds in (2.23). Nevertheless, the whole line constants $k(p,R)$ are relatively poor lower bounds for the half line constants $k(p,R^+)$. It does not seem easy to get "good" lower bounds in the half line case.

Next we discuss the question of getting upper bounds for both the whole line and the half line cases. Theorem 2.10 shows that the largest values of these constants occur for the sup norm.

Theorem 2.10 *We have*

$$(a) \quad k(p, R^+) \leq k(\infty, R^+) = 4, \quad 1 \leq p \leq \infty. \tag{2.24}$$

$$(b) \quad k(p, R) \leq k(\infty, R) = 2, \quad 1 \leq p \leq \infty. \tag{2.25}$$

Proof. The equalities were established in Section 2. The proof of the inequalities can be found in Certain and Kurtz [1977]. See also Ljubic [1964], Gindler and Goldstein [1975]. □

Lemma 2.3 *If* $3/2 < p < 2$, *then*

$$k(p, R) \leq 1/(p - 1). \tag{2.26}$$

Proof. We use Theorem 2.3 with M_4. Let $y \in M_4$. Integration by parts and Hölder's inequality gives

$$\left| \int_0^1 y'^2 y^{p-2} \right| = \left| \int_0^1 y' y^{p-2} y' \right| = \left| (p-1)^{-1} \int_0^1 y^{p-1} y'' \right|$$
$$\leq (p-1)^{-1} \|y\|_p^{p-1} \|y''\|_p. \tag{2.27}$$

Also by Hölder's inequality

$$\int_0^1 y'^p = \int_0^1 (y'^2 y^{p-2})^{p/2} y^{p(2-p)/2}$$
$$\leq \left(\int_0^1 y'^2 y^{p-2} \right)^{p/2} \left(\int_0^1 y^p \right)^{(2-p)/2}. \tag{2.28}$$

Substituting (2.27) into (2.28) we get

$$\int_0^1 y'^p \leq (p-1)^{-p/2} \|y\|_p^{p/2} \|y''\|_p^{p/2}.$$

Hence $G(y) \leq (p-1)^{-1}$ and (2.26) follows. □

Theorem 2.11 *The best constant $k(p, R) = K^2(p, R)$ in inequality (2.4) is bounded above by $U(p)$ given by*

$$U(p) = 2, \quad 1 \le p \le 3/2, \tag{2.29}$$

$$U(p) = (p-1)^{-1}, \quad 3/2 \le p \le 2, \tag{2.30}$$

$$U(p) = (q-1)^{(2-q^n)q^{-n}} \left(\prod_{i=1}^{n} (q/(q^i - 1) - 1)^{q-i} \right)^{2(q-1)},$$

$$2 < p < \infty, \quad p^{-1} + q^{-1} = 1, \tag{2.31}$$

where n is the integer

$$n = [(\log_2 q)^{-1}]. \tag{2.32}$$

Here $[\cdot]$ denotes the greatest integer function; the logarithm is taken to the base 2. Note that $n = 0$ if $1 < p < 2$, $n \ge 1$ if $2 \le p < \infty$. Further note that $q^k \le 2$ for all $k \in \{0, 1, \ldots, n\}$, but $q^{n+1} > 2$.

Equality (2.29) is a restatement of (2.25) of Theorem 2.10, and (2.30) is a restatement of Lemma 2.3.

The proof of Theorem 2.11 is long. It depends on Theorem 2.3 and on several lemmas.

Lemma 2.4 *Let k be any integer from the set $\{0, 1, \ldots, n\}$. Then*

$$\int f'^p \le C_p(k) \left(\int f^{p(q^k - 1)} f'^{p(2 - q^k)} \right)^{q^{-k}} \left(\int |f''|^p \right)^{1 - q^{-k}} \tag{2.33}$$

for any $f \in M_4$, where $C_p(0) = 1$ and

$$C_p(k) = \prod_{i=1}^{k} \left(\frac{q}{q^i - 1} - 1 \right)^{q^{-i+1}}, \quad k = 1, 2, \ldots, n. \tag{2.34}$$

Proof. The proof is by induction on k.

For $k = 0$, the right member of the inequality (2.33) reduces to $C_p(0) \int f'^p$. Because $C_p(0) = 1$, the lemma holds for $k = 0$.

Assume (2.33) holds for $k = l$. Integrating by parts and applying Hölder's inequality, we obtain the estimate

$$\int f^{p(q^l - 1)} f'^{p(2 - q^l)} = (f^{p(q^l - 1)} f') f'^{p(2 - q^l) - 1}$$

$$- \frac{p(2 - q^l) - 1}{p(q^l - 1) + 1} \int f^{p(q^l - 1) + 1} f'^{p(2 - q^l) - 2} (-f'')$$

$$\le \frac{p(2 - q^l) - 1}{p(q^l - 1) + 1} \left(\int f^{(p(q^l - 1) + 1)q} f'^{(p(2 - q^l) - 2)q} \right)^{q-1} \left(\int |f''|^p \right)^{p-1}.$$

But $(p(2-q^l)-1)(p(q^l-1)+1)^{-1} = q(q^{l+1}-1)^{-1}-1$, $(p(q^l-1)+1)q = p(q^{l+1}-1)$, and $(p(1-q^l)-2)q = p(2-q^{l+1})$, so the estimate reduces to

$$\int f^{p(q^l-1)} f'^{p(2-q^l)} \le \left(\frac{q}{q^{l+1}-1}-1\right) \left(\int f^{p(q^{l+1}-1)} f'^{p(2-q^{l+1})}\right)^{q-1} \left(\int |f''|^p\right)^{p-1}.$$

Substituting this estimate into the inequality (2.33) for $k = l$, we find

$$\int f'^p \le \left(\frac{q}{q^{l+1}-1}-1\right)^{q-l} C_p(l)$$
$$\times \left(\int f^{p(q^{l+1}-1)} f'^{p(2-q^{l+1})}\right)^{q-(l+1)} \left(\int |f''|^p\right)^{1-q^{-(l+1)}}.$$

The coefficient in the right member equals $C_p(l+1)$. We therefore conclude that (2.33) holds for $k = l+1$.

The process of induction can be continued as long as $2-q^k$ remains nonnegative, i.e., for $k = 0, 1, \ldots, n$. Thus Lemma 2.4 is proved. $\square$

To estimate $\int f'^p$ further in terms of $\int f^p$ and $\int |f''|^p$, we shall use the inequality (2.33) for $k = n$ and establish an estimate for the first integral in the right member of that inequality.

Lemma 2.5 *For any $f \in M_4$ we have*

$$\left(\int f^{p(q^n-1)} f'^{p(2-q^n)}\right)^{q-n} \le C'_p(n) \left(\int f^p\right)^{1/2} \left(\int |f''|^p\right)^{1/2(2-q^n)q^{-n}}, \qquad (2.35)$$

where the positive constant $C'_p(n)$ is given by

$$C'_p(n) = (q-1)^{(1/2)p(2-q^n)q^{-n}}. \qquad (2.36)$$

Here n is the nonnegative integer defined in (2.32).

Proof. We use Hölder's inequality:

$$\int f^{p(q^n-1)} f'^{p(2-q^n)} = (f^{p-2} f'^2)^{p(2-q^n)/2}$$
$$= \int (f^{p-2} f'^2)^{p(2-q^n)/2} (f^p)^{1-p(2-q^n)/2}$$
$$\le \left(\int f^{p-2} f'^2\right)^{p(2-q^n)/2} \left(\int f^p\right)^{1-p(2-q^n)/2}.$$

The first integral in the last expression can be evaluated by partial integration and then estimated by Hölder's inequality:

$$\int f^{p-2} f'^2 = \int (f^{p-2} f') f' = -\frac{1}{p-1} \int f^{p-1} f'' \le \frac{1}{p-1} \left(\int f^p\right)^{q-1} \left(\int |f''|^p\right)^{p-1}.$$

The factor $(p-1)^{-1}$ is equal to $(q-1)$. Thus because $(1/2)p(2-q^n)q^{-1} + 1 - (1/2)p(2-q^n) = (1/2)q^n$, it follows that

$$\int f^{p(q^n-1)} f'^{p(2-q^n)} \leq (q-1)^{p(2-q^n)/2} \left(\int f^p \right)^{q^n/2} \left(\int |f''|^p \right)^{(2-q^n)/2}.$$

The inequality (2.35) follows upon taking the q^{-n}-th power. □

Proof. To prove Theorem 2.11, we take $k = n$ in Lemma 2.4 and use (2.35), (2.36) of Lemma 2.5. This gives the inequality

$$\int f'^p \leq C_p(n)C'_p(n) \left(\int f^p \right)^{1/2} \left(\int |f''|^p \right)^{1/2}$$

which holds for any $f \in M_4$. Taking the $2p^{-1}$-th power yields

$$\frac{\|f'\|^2}{\|f\|_p \|f''\|_p} \leq (C_p(n)C'_p(n))^{2p^{-1}}, \qquad f \in M_4.$$

It follows that the quantity $k(p, M_4)$ defined in Section 4 is bounded above by the constant $(C_p(n)C'_p(n))^{2p^{-1}}$. Theorem 2.3 then implies that this same constant is therefore an upper bound for the best constant in Landau's inequality, i.e., $k(p, R)$. □

Numerical Results. For $p = 3, 4, 5$, and 6, the values of $U(p)$ are the same as those found in Kwong and Zettl [1979a].

$$p = 3: \qquad q = \frac{3}{2},\ n = 1,$$
$$U(p) = 2^{1/3} = 1.25992\,105;$$
$$p = 4: \qquad q = \frac{4}{43},\ n = 2,$$
$$U(p) = (15/7)^{3/8} = 1.33082\,962;$$
$$p = 5: \qquad q = \frac{5}{4},\ n = 3,$$
$$U(p) = 4^{47/125}(11/9)^{8/25}(19/61)^{32/125} = 1.33222\,966;$$
$$p = 6: \qquad q = \frac{6}{5},\ n = 3,$$
$$U(p) = 5^{19/108}(19/11)^{5/18}(59/91)^{25/108} = 1.39745\,611.$$

(Note that in Kwong and Zettl [1979a] the exponent of $(19/11)$ in $U(6)$ is erroneously given as $5/8$ instead of $5/18$.)

Table 2.1 lists the approximate values of the lower bounds $L(p)$ from Theorem 2.7 and the upper bounds $U(p)$ from Theorem 2.11 for various values of p. For comparison purposes we also list the numerical values of $2^{(p-2)/p}$ to 5 decimal places. These were conjectured to be the best constants $k(p, R)$ in Gindler and Goldstein [1975].

Table 2.1: Numerical Values of $L(p)$, $U(p)$, and $2^{1-2/p}$

p	$L(p)$	$U(p)$	$2^{1-2/p}$	p	$L(p)$	$U(p)$	2^{1-2p}
2.0	.912871	1.000000	1.000000	70.0	1.828515	1.877338	1.960781
2.5	.975860	1.275425	1.148698	80.0	1.844562	1.888009	1.965641
3.0	1.030321	1.259921	1.259921	90.0	1.857614	1.896313	1.969430
3.5	1.078026	1.204980	1.345900	100.0	1.868562	1.904110	1.972465
4.0	1.20263	1.330830	1.414214	200.0	1.923167	1.946654	1.986185
4.5	1.158000	1.326950	1.469734	300.0	1.944516	1.961656	1.990779
5.0	1.191978	1.332230	1.515717	400.0	1.956136	1.969043	1.993081
5.5	1.222775	1.405542	1.554406	500.0	1.963519	1.974519	1.994463
6.0	1.250854	1.397456	1.587401	600.0	1.968857	1.978507	1.995384
6.5	1.276586	1.420297	1.615866	700.0	1.972454	1.980776	1.996043
7.0	1.300276	1.466459	1.640671	800.0	1.975382	1.982992	1.996537
7.5	1.322176	1.454720	1.662476	900.0	1.977715	1.984773	1.996922
8.0	1.342495	1.485189	1.681793	1000.0	1.979620	1.985880	1.997229
8.5	1.361413	1.514963	1.699024	2000.0	1.988750	1.992150	1.998614
9.0	1.379078	1.500704	1.714488	3000.0	1.992089	1.994537	1.999076
9.5	1.395620	1.535140	1.728444	4000.0	1.993849	1.995804	1.999307
10.0	1.411150	1.554116	1.741101	5000.0	1.994944	1.996568	1.999446
11.0	1.439549	1.574849	1.763183	6000.0	1.995695	1.997076	1.999538
12.0	1.464908	1.569207	1.781797	7000.0	1.996244	1.997438	1.999604
13.0	1.487720	1.613182	1.797702	8000.0	1.996663	1.997707	1.999653
14.0	1.508372	1.634090	1.811447	9000.0	1.996994	1.997915	1.999692
15.0	1.527175	1.628411	1.823445	10000.0	1.997263	1.998124	1.999723
16.0	1.544380	1.655550	1.834008	20000.0	1.998527	1.99001	1.999861
17.0	1.560195	1.676200	1.843379	30000.0	1.998977	1.999298	1.999908
18.0	1.574790	1.676796	1.851749	40000.0	1.999211	1.999466	1.999931
19.0	1.588409	1.687403	1.859271	50000.0	1.999356	1.999556	1.999945
20.0	1.600874	1.707650	1.866066	60000.0	1.999454	1.999631	1.999954
30.0	1.691033	1.777383	1.909683	70000.0	1.999525	1.999674	1.999960
40.0	1.745134	1.817898	1.931873	80000.0	1.999580	1.999716	1.999965
50.0	1.781684	1.844533	1.945310	90000.0	1.999622	1.999741	1.999969
60.0	1.808240	1.863373	1.954320	100000.0	1.999657	1.999768	1.999972

Franco, Kaper, Kwong and Zettl [1983]

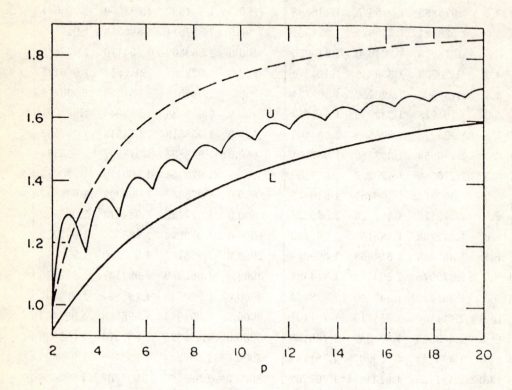

Figure 2.1: Graph of the lower (L) and upper (U) bounds of the best constants in Landau's inequality on the real line. The dashed curve represents the graph of the function $p \to 2^{1-2/p}$.

2.8 Extremals

In this section we discuss the existence and nature of extremals for inequality (2.4) in the whole line case R and the half-line case R^+ as well as for a number of closely related finite interval problems. These include some of the problems of Sections 4 and 5. There we introduced problems M_i, $i = 1, \ldots, 9$ which are related to the whole-line case and problems N_1 and N_2 associated with the half-line case.

Since the two extreme cases $p = 1$ and $p = \infty$ were discussed in Sections 4 and 5, we assume $1 < p < \infty$ for the rest of this section.

In this section we prove that for all values of p, $1 < p < \infty$, inequality (2.4) has an extremal in the half-line case R^+ and not in the whole-line case R. Both our existence and nonexistence proofs are based on Kwong and Zettl [1980a]. As far as we know the proof given there is the only known proof of the existence of extremals for inequality (2.4) without the explicit knowledge of the constant $K = K(p, J)$. In all cases where extremals were previously known to exist, their existence was established by first proving (2.4) for some explicit value of K and then giving an ad hoc construction of a nontrivial function for which equality holds.

Here an existence proof is given valid for all p, $1 < p < \infty$ and $J = R^+$. This proof is long and technical. The basic idea, however, is simple: Show that G defined by (2.8) is a continuous function on a compact set S in some appropriate function space. To carry out this procedure we must overcome two obstacles:

1. Since G "blows up" at $y = 0$ we must show that extremals cannot be arbitrarily small.

2. Somehow a compact set S must be found that contains extremals. The set of concave functions and the bounded interval characterizations of Section 2.5 will play an important role in this.

Before proceeding to the main results, several lemmas are established. Some of these are of independent interest.

Lemma 2.6 *Let* $1 < p < \infty$. *Suppose* $f \in M_4$. *If* f *is not strictly increasing or concave, then there exists a* $g \in M_4$ *such that* $G(g) > G(f)$ *and* $\|g''\|_p \leq \|f''\|_p$.

Proof. By definition, functions in M_4 are monotonically increasing, but need not be strictly monotone. Suppose first that $f \in M_4$ is not strictly monotone. Then there are points

$t_1 < t_2$ in $[0,1]$ such that $f(t_1) = f(t_2)$. Define

$$g(t) = \begin{cases} f(t - t_2 + t_1) & t \in [t_2 - t_1, t_2] \\ f(t) & t \in [t_2, 1] \end{cases}.$$

After scaling, $g(t)$ belongs to M_4. Obviously $\|g\|_p < \|f\|_p$, $\|g'\|_p = \|f'\|_p$, and $\|g''\|_p = \|f''\|_p$.
Hence $G(g) > G(f)$ and the lemma is proved.

We thus assume now that f is strictly monotone. ¿From the hypothesis, f' is not monotonically decreasing. There are thus points $t_1 < t_2$ in $[0,1]$ such that $f'(t_1) < f'(t_2)$. In
$[t_1, t_2]$ choose a point t_3 at which f' attains its maximum. The point t_3 may coincide with
t_2 but not t_1. Thus the interval $[t_1, t_3]$ is nondegenerate and $f'(t) \le f'(t_3)$ for all $t \in [t_1, t_3]$.
Define a continuous function h on $[0,1]$ by: $h(t) = f(t)$ in $[t_3, 1]$, $h'(t) = f'(t_3)$ in $[t_1, t_3]$
and $h'(t) = f'(t)$ in $[0, t_1]$. Then $h(t) < f(t)$ for $t \in [0, t_1]$. Thus h has a zero, say α,
in $(0,1)$. Since f and h are strictly monotone functions, their inverses f^{-1} and h^{-1} exist.
Clearly $h^{-1}(x) \ge f^{-1}(x)$ for all x in $[0, f(1)]$, while $h^{-1}(x) > f^{-1}(x)$ for $x \in [0, f(t_1)]$.
Let $u = f' \circ f^{-1}$ and $v = h' \circ h^{-1}$, where the circle denotes composition. In $[f(t_3), f(1)]$,
$u(x) = v(x)$ since f coincides with h. In $[f(t_1), f(t_3)]$,

$$u(x) \le f'(t_3) = v(x) \text{ while } u(f(t_1)) < v(f(t_1)). \tag{2.37}$$

Suppose $u(x) \le v(x)$ for all $x \in [0, f(t_1)]$. Then by a change of variable

$$\int_0^1 f'^p(t)dt = \int_0^{f(1)} u^{p-1}(x)dx < \int_0^{f(1)} v^{p-1}(x)dx$$

$$= \int_\alpha^1 v'^p(t)dt.$$

On the other hand,

$$\int_\alpha^1 h^p(t)dt \le \int_\alpha^1 f^p dt < \int_0^1 f^p(t)dt,$$

and

$$\int_\alpha^1 |h''|^p = \int_\alpha^{x_1} |h''|^p + \int_{x_3}^1 |h''|^p = \int_\alpha^{x_1} |f''|^p + \int_{x_3}^1 |f''|^p < \int_0^1 |f''|^p.$$

Thus $G(f) < G(h_\alpha)$, where h_α is the restriction of h to $[\alpha, 1]$. By scaling we get a function
g in M_4 from h_α.

Suppose now that for some point x, $u(x) > v(x)$. Let x_1 be the supremum of such
points in $[0,1]$. By continuity $u(x_1) = v(x_1)$. By (2.37) we see that $x_1 < f(t_1)$, and hence
$f^{-1}(x_1) < h^{-1}(x_1)$. Now we translate that part of the graph of f below $x = x_1$ to the right
until it joins the graph of h. This translated portion along with that part of the curve of h
above $x = x_1$ forms a new smooth curve that defines a function g which, after scaling, can be
regarded as in M_4. Let $w = g' \circ g^{-1}$. Then $g(t) \le f(t)$ and $w(x) \le u(x)$. The same analysis
used above shows that $G(g) > G(f)$. This completes the proof of Lemma 2.6. $\square$

In particular, the quotient $G(y)$ and hence the constant $k(p, M)$ is independent of the compact interval $J = [a, b]$. That is, if y is defined on some compact interval $[a, b]$, we may, by horizontal scaling and translation, assume that y is defined on $[0, 1]$ (or any other compact interval), i.e., we may replace y by a function z defined on $[0, 1]$ such that $G(y) = G(z)$. This observation will be used often and repeatedly below, sometimes without explicit mention.

We consider the following sets of functions for $1 \le p \le \infty$:

$$M_1 = \{y \in W_p^2([0,1]) : y(0) = y(1) = 0\}$$
$$M_2 = \{y \in M_1 : y(t) > 0, 0 < t < 1\}$$
$$M_3 = \{y \in W_p^2([0,1]) : y(0) = y'(1) = 0, y(t) > 0, 0 < t < 1\}$$
$$M_4 = \{y \in M_3 : y'(t) \ge 0, 0 < t < 1\}$$
$$M_5 = \{y \in W_p^2([0,1]) : y'(0) = 0 = y'(1)\}$$
$$M_6 = \{y \in W_p^2(R^+) : y(0) = 0\}$$
$$M_7 = \{y \in W_p^2(R^+) : y'(0) = 0\}$$
$$M_8 = \{y \in W_p^2(R^+) : y'(0)y(0) \ge 0\}$$
$$M_9 = \{y \in W_p([0,1]) : y'(1) = 0, y'(t) \ge 0, y(t) \ge 0, 0 < t < 1\}$$

Theorem 2.3 *For any* p, $1 \le p < \infty$,

$$k(p, M_i) = k(p, R), \quad i = 1, 2, \ldots, 9. \tag{2.9}$$

Furthermore, the interval $[0, 1]$ *in the definition of* M_i, $i = 1, 2, \ldots, 9$ *can be replaced by any compact interval* $[a, b]$ *and the end point conditions transferred accordingly.*

Proof. The "furthermore" statement follows from the observations made in the paragraph preceding the definition of the M_i.

The proof of (2.9) is based on two lemmas. The first one is a special case of Lemma 1.6 of section 3 in Chapter 1, and is stated here only for the convenience of the reader. □

Lemma 2.1 *Let* $y \in W_p^2(R)$, $1 \le p < \infty$. *For any* $\epsilon > 0$ *there exists a* $z \in C_0^\infty(R)$ *such that*

$$\|y - z\|_p < \epsilon, \quad \|y' - z'\|_p < \epsilon, \quad \|y'' - z''\|_p < \epsilon. \tag{2.10}$$

Proof. See Lemma 1.6 of Section 3 in Chapter 1. □

The next lemma plays a fundamental role in this section and several subsequent ones. It involves the W_p^2 spaces over unions of intervals I_i. By

$$y \in W_p^2(\cup_{i=1}^n I_i)$$

we mean that the restriction of y to I_i is in $W_p^2(I_i)$ for each $i = 1, \ldots, n$.

2.7 Upper and Lower Bounds for $k(p, R)$ and $k(p, R^+)$

Since the exact values of these constants are known only for $p = 1, 2, \infty$, we consider the problem of obtaining "good" upper and lower bounds. To do this it is useful to use some of the compact interval characterizations of these constants developed in Sections 4 and 5.

A lower bound can be obtained simply by evaluating $G(y)$ for a particular y from the appropriate set of functions.

Theorem 2.6 *For any p, $1 \leq p \leq \infty$, we have*

$$1 \leq k(p, R). \tag{2.19}$$

Proof. Consider $y(t) = \sin(\pi t/2)$, $t \in [0, 2]$. Then $y \in M_1$ when the interval $[0,1]$ is replaced by $[0,2]$. A calculation shows that $G(y) = 1$ and (2.19) follows from Theorem 2.3. $\square$

Theorem 2.7 *For any p, $1 \leq p < \infty$, we have*

$$k^p(p, R) \geq 2^{p+1}\Gamma(p + 1 + 1/2)/[\pi^{1/2}\Gamma(p + 1)\Gamma(p + 2)] = (L(p))^p, \tag{2.20}$$

where Γ denotes the gamma function.

Proof. Consider

$$y(t) = t(1 - t), \quad t \in [0, 2].$$

Then y is in M_2 on the interval $[0,2]$. A calculation yields

$$\int_0^2 y^p = 2^{2p+1}\Gamma^2(p + 1)/\Gamma(2p + 2) = \sqrt{\pi}\Gamma(p + 1)/\Gamma(p + 1 + 1/2)$$
$$\int_0^2 |y'|^p = 2^{p+1}/(p + 1)$$
$$\int_0^2 |y''|^p = 2^{p+1}.$$

Thus $G(y)$ is the right-hand side of (2.20) and the conclusion follows from Theorem 2.3. $\square$

We will show in Section 9 that $k(p, R)$ is a continuous function of p. Since $k(2, R) = 1$, $k(\infty, R) = 2$, and 2 is the limit as $p \to \infty$ of the right-hand side of (2.20), it follows that (2.20) is a "good" lower bound when p is large and 1 is a good lower bound when p is near 2.

The next result improves on (2.19). The idea is to use a linear combination of the two functions employed in the proofs of Theorems 2.6 and 2.7 to improve the lower bound. Although this does not yield an explicit expression in terms of p it does show that strict inequality holds in (2.19) when $p \neq 2$.

constant that depends only on I and not on the particular partition of I. This can be done if we define

$$B(p, k, w, I) = \inf_{\underline{I}} B(p, k, w, \underline{I})$$

where the infimum is taken over all partitions satisfying the hypotheses of Lemma 3.1. Thus (3.12) can be given as

$$\alpha_k^p \leq B(p, k, w, I) \int_I |f|^p w. \tag{3.12'}$$

Theorem 3.2 *Suppose the hypotheses of Theorem 3.1 hold and $1 \leq p < \infty$. Let n be a positive integer ≥ 1. Then for any admissible function f, we have*

$$\int_I |f^{(n-1)}|^p v \leq A \int_I |f^{(n)}|^p u + B^* \int_I |f|^p w, \tag{3.19}$$

where $A = A(p, u, v, I)$ is given by (3.9) (with the modification indicated there when $p = 1$) and

$$B^* = B^*(p, n-1, v, w, I) = 2^{p-1} \left(\int_I v \right) B(p, n-1, w, \underline{I})$$

with B given by (3.13).

Proof. Set $\alpha_{n-1} = \inf_I |f^{(n-1)}(t)|$. Applying (3.11) to $f^{(n-1)}$ and then using Lemma 3.1 we get

$$
\begin{aligned}
|f^{(n-1)}(t)|^p &\leq 2^{p-1} \left[\alpha_{n-1}^p + \left(\int_I u^{-1/p} \right)^{p/q} \int_I |f^{(n)}|^p u \right] \\
&\leq 2^{p-1} \left[B \int_I |f|^p w + \left(\int_I u^{-q/p} \right)^{p/q} \int_I |f^{(n)}|^p u \right]. \tag{3.20}
\end{aligned}
$$

Multiplying (3.20) by v and integrating over I yields (1.19). $\square$

Remark 3.3 Note that the constant A in (3.19) depends only on p, u, v, and I; B^* only on p, n, v, w, and I.

In the general case when the left-hand side of (3.19) involves lower order derivatives, we obtain a weaker result.

Theorem 3.3 *Suppose the hypotheses of Theorem 3.1 hold, $1 \leq p < \infty$, and I has finite length L. Let k and n be integers satisfying $0 \leq k < n$. Then for any admissible function f*

$$\int_I |f^{(k)}|^p v \leq C \int_I |f^{(n)}|^p u + D \int_I |f|^p w, \tag{3.21}$$

where

$$C = C(p,k,n,u,v,I) = 2^{(\alpha+1)(p-1)}L^{\alpha p}\left(\int_I u^{-q/p}\right)^{p/q}\int_I v, \quad 1 < p < \infty$$

$$= L^\alpha(\sup_I u^{-1})\int_I v, \quad p = 1 \tag{3.22}$$

$$D = D(p,k,n,v,w,I) = 2^{p-1}\int_I v[2^{\alpha(p-1)}L^{\alpha p}B(p,n-1,w,\underline{I})$$

$$2^{(\alpha-1)(p-1)}L^{(\alpha-1)p}B(p,n-2,w,\underline{I}) + \ldots +$$

$$2^{p-1}L^p B(p,k+1,w,\underline{I})$$

$$+ B(p,k,w,\underline{I})] \tag{3.23}$$

with $\alpha = n - k - 1$.

Furthermore, if $w > 0$ almost everywhere on I, then the constant D can be chosen as

$$D = D(p,k,n,v,w,I) = 2^{p-1}\left(\int_I v\right)L^{-pk}M_{n-1}^{-1}[2^{\alpha(p-1)}E(p,n-1)$$

$$+ 2^{(\alpha-1)(p-1)}E(p,n-2) + \ldots + 2^{p-1}E(p,k+1) + E(p,k)], \tag{3.24}$$

where

$$E(p,r) = 2^{r(p-1)}(2^{r+1}-1)^{pr}, \quad r = 0,1,\ldots,n-1$$

and

$$M_{n-1} = \min_{0\le k \le n-1}\left(\min_{J\in I^{(k)}}\int_J w\right),$$

where

$$I^{(k)} = \{I_i^{(k)}, \ i = 1,3,5,\ldots,2^{k+1}-1\}, \quad 0 \le k \le n-1$$

and for each k, $0 \le k \le n-1$, the min $J \in I^{(k)}$ is taken over all intervals $J = I_i^{(k)}$, $i = 1,3,5,\ldots,2^k - 1 = m_k$, where $I_1^k = [a_1 + \epsilon, b_1]$, $I_{m_k} = [a_{m_k}, b_{m_k} - \epsilon]$ for $\epsilon = L/(2m_k)$ and $I_1^{(k)} = J_i$, $i = 2,3,\ldots,m_k - 1$ are such that $J_i = (a_i,b_i)$, $i = 1,\ldots,m_k$ form a partition of the interval $I = (a,b)$ into subintervals of equal length.

Proof. Let

$$A = A(p,u,v,I) = 2^{p-1}\int_I v\left(\int_I u^{-q/p}\right)^{p/q}, \quad p^{-1} + q^{-1} = 1, \tag{3.25}$$

modified as in (3.9) when $p = 1$,

$$B^* = B^*(p,k,v,w,I) = 2^{p-1}B\int_I v, \tag{3.26}$$

where B is given by (3.13).

The proof of (3.21) will be by a backward induction on k. By Theorem 3.2, (3.21) holds for $k = n - 1$. Suppose that (3.21) holds for some k, $0 < k < n$. Then by Theorem 3.2 with $n = k$ and $u = 1$, we have

$$\int_I |f^{(k-1)}|^p v \leq A(p, 1, v, I) \int_I |f^{(k)}|^p . 1 + B^*(p, k-1, v, w, I) \int_I |f|^p w. \tag{3.27}$$

Applying the induction hypothesis (1.21) with $v = 1$ we get

$$\int_I |f^{(k)}|^p . 1 \leq C(p, k, n, u, 1, I) \int_I |f^{(n)}|^p u + D(p, k, n, 1, w, I) \int_I |f|^p w. \tag{3.28}$$

Using (3.28) in (3.27) we obtain

$$\begin{aligned}
\int_I |f^{(k-1)}|^p v \leq \ & A(p, 1, v, I) \Big[C(p, k, n, u, 1, I) \int_I |f^{(n)}|^p u \\
& + D(p, k, n, 1, w, I) \int_I |f|^p w \Big] \\
& + B^*(p, k-1, v, w, I) \int_I |f|^p w.
\end{aligned} \tag{3.29}$$

From (3.22) and (3.25)

$$A(p, 1, v, I)\, C(p, k, n, u, 1, I) = C(p, k-1, n, u, v, I).$$

From (3.23), (3.25), and (3.26)

$$A(p, 1, v, I)\, D(p, k, n, 1, w, I) + B^*(p, k-1, v, w, I) = D(p, k-1, n, v, w, 1).$$

Hence (3.21) holds with k replaced by $k - 1$ and the proof of the first part of Theorem 3.3 is complete. To prove the furthermore part we replace (3.13) by (3.24) and proceed as above. □

Theorem 3.4 *Let I be a bounded open interval, let k and n be integers with $0 \leq k < n$, let $1 \leq p < \infty$ and $p^{-1} + q^{-1} = 1$. Assume $u \geq 0$, $v \geq 0$, $w > 0$ almost everywhere on I, $u^{-q/p} \in L^1(I)$ ($u^{-1} \in L^\infty(I)$ when $p = 1$), $v \in L^1(I)$. Then, for any $\epsilon > 0$, there exists a $K(\epsilon) > 0$ (depending only on ϵ, I, k, n, p, u, v, and w) such that for all admissible f*

$$\int_I |f^{(k)}|^p v \leq \epsilon \int_I |f^{(n)}|^p u + K(\epsilon) \int_I |f|^p w. \tag{3.30}$$

Proof. Let $\epsilon > 0$. From (3.22) and the absolute continuity of the integral, there exists a $\delta > 0$ such that $C(p, k, n, u, v, J) < \epsilon$ for any subinterval J of I of length $< \epsilon$. Let $I = \cup_{i=1}^m J$ be a partition of I into non-overlapping subintervals each of length $< \epsilon$. By Theorem 3.3, (3.21) holds for $I = J_i$ with $C = C(p, k, n, u, v, J_i) < \epsilon$ and $D = D(p, k, n, v, w, J_i)$, $i = 1, 2, \ldots, m$. By summing all these inequalities over i we get (3.30) with

$$K(\epsilon) = \max D(p, k, n, v, w, J_i) \text{ for } i = 1, 2, \ldots, m. \qquad □$$

Corollary 3.1 *For each $\epsilon > 0$ there exists $K_1(\epsilon) > 0$ such that for all admissible functions f we have*

$$\|f^{(k)}\|_v \leq \epsilon \|f^{(n)}\|_u + K_1(\epsilon)\|f\|_w, \tag{3.31}$$

where $K_1(\epsilon) = K^{1/p}\epsilon^p$ with $K = K(\epsilon)$ as in (3.30).

Proof. The inequality (3.31) follows from (3.30) and the elementary inequality (3.5). □

Remark 3.4 The special case $k = 0$, $u = 1$, $p = 2$ of (3.31) was established by Everitt [3] who also showed by examples that, in this case, $K(\epsilon)$ may or may not be uniformly bounded as $\epsilon \to 0$. By taking the limit as $\epsilon \to 0$ we see that the uniform boundedness of $K(\epsilon)$ is equivalent to the fact that the differentiation operator $(d/dt)^k$ is a bounded operator from $L^p(I, w)$ to $L^p(I, v)$. It follows that for $k > 0$ we always have $\lim_{\epsilon \to 0} K(\epsilon) = \infty$, except in the trivial case when $v = 0$ almost everywhere. In case $k = 0$, if $v \geq Kw$ almost everywhere then $\int_I |f|^p v \leq K \int_I |f|^p w + \epsilon \int_I |f'|^p u$ for all admissible f and all $\epsilon > 0$. On the other hand, if $\int_I |f|^p v \leq \epsilon \int_I |f'|^p u + K \int_I |f|^p w$ holds for all $\epsilon > 0$ and all admissible f, then letting $\epsilon \to 0$ we conclude that $\int_I |f|^p v \leq K \int_I |f|^p w$ for all admissible f. Hence $v \leq Kw$ almost everywhere. Thus we conclude that in the case $k = 0$ the constant $K(\epsilon)$ can be chosen uniformly bounded as $\epsilon \to 0$ if and only if $v \leq Kw$ almost everywhere for some constant K.

In the case of an unbounded interval J the conditions (3.6) on the coefficient functions are rather restrictive. Now we seek to relax these. But to get an inequality of type (3.2) on an unbounded interval without conditions (3.6), we need some stringent growth restrictions on u, v, w. Furthermore, these restrictions are not independent of each other.

Theorem 3.5 *Let I be an open bounded interval, let $1 \leq p < \infty$, let α be any real number, and let n, k be integers with $0 \leq k < n$.*

Suppose s and w are positive functions which are absolutely continuous on all compact subintervals of I and

$$(i) \quad |s'(t)| \leq N \qquad \qquad almost\ everywhere\ on\ I$$
$$(ii) \quad |s(t)w'(t)| \leq Mw(t) \quad almost\ everywhere\ on\ I$$

for some positive constants N and M. Then for any $\epsilon > 0$ there exists a constant $K(\epsilon) > 0$, depending on ϵ, s, w, α, p, n, k, and I such that for all admissible functions f,

$$\|s^{\alpha+k}f^{(k)}\|_w^p \leq \epsilon \|s^{\alpha+n}f^{(n)}\|_w^p + K(\epsilon) \|s^{\alpha}f\|_w^p. \tag{3.32}$$

Furthermore, if

$$\int_I s^{-1} = \infty, \tag{3.33}$$

then $K(\epsilon) = K(\epsilon, N, M, \alpha, p, n, k)$ is independent of I, s, and w but depends on the constants N and M as well as on ϵ, α, p, n, and k.

The proof is long and technical and will be given with the help of several lemmas.

Lemma 3.2 *For any $\epsilon > 0$ there exists a constant $\bar{K}(\epsilon) > 0$ such that if $I = (a,b)$ is any bounded interval of length L, then*

$$\|f'\| \leq \epsilon L^p \|f''\|^p + \bar{K}(\epsilon) L^{-1} \|f\|^p \tag{3.34}$$

for all admissible functions f. Here $\bar{K}(\epsilon) = \bar{K}(\epsilon, p)$ is independent of f and I.

Proof. By (3.30) with $u = v = w = 1$, $n = 2$, $k = 1$, and $I = (0,1)$ we have

$$\int_0^1 |g'|^p \leq \epsilon \int_0^1 |g''|^p + \bar{K}(\epsilon) \int_0^1 |g|^p. \tag{3.35}$$

Taking $g(t) = f(Lt + a)$ in (3.35) we obtain (3.34). $\square$

Lemma 3.3 *Assume that (i) of Theorem 3.5 holds and I is a bounded interval of length L. If*

$$\int_I s^{-1} = (2N)^{-1}, \tag{3.36}$$

then

$$\sup_I s(t) \leq 3 \inf_I s(t) \tag{3.37}$$

and

$$L \sup_I s^{-1}(t) \leq N^{-1}, \quad L^{-1} \sup_I s(t) \leq 3N. \tag{3.38}$$

Proof. Let $\alpha = \sup s(t)$ for t in I, $\beta = \inf s(t)$ for t in I. From (i) we have for any t_1, t_2 in I

$$|s(t_2) - s(t_1)| \leq N|t_2 - t_1| \leq NL. \tag{3.39}$$

Hence

$$\alpha - \beta \leq NL. \tag{3.40}$$

Also,

$$L/\alpha = \int_I \alpha^{-1} \leq \int_I s^{-1} = (2N)^{-1}. \tag{3.41}$$

From (3.40) and (3.41)

$$\beta \geq \alpha - NL \geq NL > 0. \tag{3.42}$$

Also,

$$(2N)^{-1} = \int_I s^{-1} \leq L/\beta. \tag{3.43}$$

From (3.40), (3.41), and (3.43) follows

$$2NL \leq \alpha \leq 3NL. \tag{3.44}$$

Now (3.37) and (3.38) are consequences of (3.42), (3.43), and (3.44). $\square$

Lemma 3.4 *Let I be bounded of length L. Assume that (1.36) and (ii) of Theorem 3.5 hold. Then*

$$\sup_I w(t) \leq \exp(M/2N) \inf_I w(t). \tag{3.45}$$

Proof. ¿From (ii) and (3.36) we have, for any points t_1 and t_2 of I,

$$|\log(w(t_2)/w(t_1))| = \left| \int_{t_1}^{t_2} w'/w \right| \leq M \int_{t_1}^{t_2} s^{-1} \leq M/(2N) \tag{3.46}$$

and (3.45) follows. $\square$

Proof. (of Theorem 3.5) We first establish the special case $\alpha = 0$. If $\int_I s^{-1} < \infty$ then I must be bounded since (i) implies that $s^{-1}(t) \geq [N(t - t_0) + s(t_0)]^{-1}$ for all $t \geq t_0$, t_0 fixed in I. Let $I = (a, b)$. From (3.39) and the Cauchy convergence criterion it follows that s can (if necessary) be continuously extended to $\bar{I}$, the closure on I. Similarly, from (ii) and (3.46) it follows that w can be continuously extended to $\bar{I}$. By (3.42) the extension of s is bounded away from zero on $\bar{I}$. ¿From (3.46) and (3.36) it follows that the extension of w is bounded away from zero on $\bar{I}$. Thus, for $u = s^{np}w$, $v = s^{kp}w$, we have $u^{-1} \in L^\infty(I)$ if $p = 1$ and $u^{-q/p}$, v in $L^1(I)$ if $p > 1$ and Theorem 3.5 follows from Theorem 3.4 in this case.

Assume $\int_I s^{-1} = \infty$. Let $I = \cup_{i=1}^\infty I_j$ be an infinite partition of I into nonoverlapping subintervals such that $\int_J s^{-1} = (2N)^{-1}$ for $J = I_j$, $i = 1, 2, \ldots$. Then, as above, each I_j is bounded. We first establish (3.32) on each closed bounded interval I_j for $k = 1$, $n = 2$. Fix j and let $J = I_j$. Using Lemmas 3.2, 3.3, and 3.4, we obtain

$$
\begin{aligned}
\|sf'\|_{w,J}^p &= \int_j s^p |f'|^p w \leq \sup_J s^p \sup_J w \int_J |f'|^p \\
&\leq \sup_J s^p \sup_J w \left(\epsilon_1 L^p \int_J |f''|^p + \bar{K}(\epsilon) L^{-p} \int_J |f|^p \right) \\
&\leq \exp(M/(2N)) \left[\epsilon_1 \sup_J s^{2p} \sup_J s^{-p} L^p \int_J |f''|^p w \right. \\
&\quad \left. + \bar{K}(\epsilon_1) \sup_J s^p L^{-p} \int_J |f|^p w \right] \\
&\leq \exp[M/(2N)][\epsilon_1 3^{2p} N^{-p} \|s^2 f''\|_{w,J}^p \\
&\quad + K(\epsilon_1)(3N)^p \|f\|_{w,J}^p].
\end{aligned}
\tag{3.47}
$$

By choosing ϵ_1 sufficiently small in (3.47) we may conclude that (3.32) holds on each interval I_j, $j = 1, 2, 3, \ldots$ with the same ϵ and $K(\epsilon)$. Summing over j yields (3.32) for $\alpha = 0$, $k = 1$,

$n = 2$. The case $\alpha \neq 0$ follows from the case $\alpha = 0$ by replacing w by $s^{\alpha p}w$, and noting that $s^{\alpha p}w$ satisfies (ii) if w does, provided M is replaced by $M + |a|pN$. (See Remarks below.)

The proof of the general case is by induction on n. The case $k = 1$, $n = 2$ was established above. The case $k = 0$ is trivial for any $n \geq 1$. Assume (3.32) holds for some integer $n - 1 \geq 2$ and all k, $0 \leq k < n - 1$. First consider the case $k = n - 1$. Observe that $s^p w$ satisfies (ii) with M replaced by $pN + M$. By the induction hypothesis applied to f' and the weight function $s^p w$ we have

$$\|s^{\alpha+n-2}f^{(n-1)}\|_{s^p w}^p \leq \epsilon_1 \|s^{\alpha+n-1}f^{(n)}\|_{s^p w}^p + K_1(\epsilon_1)\|s^\alpha f'\|_w^p. \tag{3.48}$$

Applying the inductive hypothesis again we obtain

$$\|s^{\alpha+1}f'\|_w^p \leq \epsilon_2 \|s^{\alpha+n-1}f^{(n-1)}\|_w^p + K_2(\epsilon_2)\|s^\alpha f\|_w^p. \tag{3.49}$$

Substituting (3.49) and (3.48) and transferring the term involving $f^{(n-1)}$ to the left-hand side yields

$$[1 - \epsilon_2 K_1(\epsilon_1)]\|s^{\alpha+n-1}f^{(n-1)}\|_w^p \leq \epsilon_1 \|s^{\alpha+n}f^{(n)}\|_w^p + K_1(\epsilon_1)K_2(\epsilon_2)\|s^\alpha f\|_w^p. \tag{3.50}$$

Given $\epsilon_3 > 0$, choose ϵ_1 so that $0 < \epsilon_1 < \epsilon_2/2$. Let $K(\epsilon_3) = 2K_1(\epsilon_1)K_2(\epsilon_2)$. Then

$$\|s^{\alpha+n-1}f^{(n-1)}\|_w^p \leq \epsilon_3 \|s^{\alpha+n}f^{(n)}\|_w^p + K_3(\epsilon_3)\|s^\alpha f\|^p w. \tag{3.51}$$

This establishes the case $k = n - 1$. Assume $0 \leq k < n - 1$. By the inductive hypothesis we have

$$\|s^{\alpha+k}f^{(k)}\|_w^p \leq \epsilon_4 \|s^{\alpha+n-1}f^{(n-1)}\|_w^p + K_4(\epsilon_4)\|s^\alpha f\|_w^p. \tag{3.52}$$

Given $\epsilon > 0$ choose the positive numbers ϵ_3, ϵ_4, so that $\epsilon_3\epsilon_4 < \epsilon$, and let $K(\epsilon) = \epsilon_4 K_3(\epsilon_3) + K_4(\epsilon_4)$. Then substituting (3.52) into (3.51) yields (3.32). This completes the proof of the main part of Theorem 3.5. The furthermore statement follows from the above proof. Since (3.47) holds on any interval I_j, satisfying $\int_J s^{-1} = (2N)^{-1}$, $j = 1, 2, 3, \ldots$, given $\epsilon > 0$ we can choose ϵ_1 in (3.47) such that

$$\exp[M/(2N)]\epsilon_1 3^{2p} N^{-p} < \epsilon$$

and then choose $K(\epsilon) = (3N)^p \bar{K}(\epsilon_1)$. Thus (3.32) follows for $k = 1$, $n = 2$ by summing over j. Hence $K(\epsilon)$ depends only on M, N, and p, not on I, w, and s as long as (i) and (ii) hold. Also, in the above induction arguments, $K(\epsilon)$ depends solely on ϵ, M, N, p, n, k, and α. $\square$

3.2 Inequalities of Product Form

In this section we study the class of weight functions w for which the inequality

$$\left(\int_J |y'|^p w\right)^{1/p} \leq K \left(\int_J |y|^p w\right)^{1/(2p)} \left(\int_J |y''|^p w\right)^{1/(2p)} \tag{3.53}$$

holds. Here $J = R$ or R^+, $1 \leq p < \infty$, K is some finite constant such that (3.53) is valid for all functions y satisfying: y' is absolutely continuous on compact subintervals of J (so that y'' exists a.e. and is locally integrable) and the two integrals on the right of (3.53) are finite. Such a y is said to be admissible. By a weight function we mean a function satisfying

$$w \geq 0, \; w \in L_{\text{loc}}(J). \tag{3.54}$$

Clearly if (3.53) holds for some finite K then there is a smallest such K. This we denote by $K = K(p, J, w)$. In the case of a unit weight function, we continue to use the notation of Chapters 1 and 2, i.e., $K(p, J, 1) = K(p, J)$.

For what weight functions w does (3.53) hold for some constant K and all admissible y? This is the question we study in the remainder of this chapter. Higher-order versions of (3.53) are also considered. Also, we try to get some information about the dependence of K on w, p, and J.

That (3.53) does not hold for general weight functions can be seen from the simple example below.

Example 1. $y(t) = t$, $w(t) = \exp(-t)$, $J = R^+ = (0, \infty)$. In this case the right-hand side of (3.53) is zero for any finite K while the left-hand side is positive.

For convenience we introduce some notation. For $0 < K < \infty$, $1 \leq p < \infty$, let $W(p, K, J)$ denote the set of all weight functions w such that (3.53) holds for all admissible y. Let

$$W(p, J) = \{W(p, K, J) : 0 < K < \infty\}.$$

The problem of finding "useful" necessary and sufficient conditions for w to be in $W(p, K, J)$ or $W(p, J)$ seems to be a rather difficult one. In the remainder of this chapter we will find a number of sufficient conditions as well as some which are necessary. However, we start by developing a general property of these weight function classes.

Theorem 3.6 *Each of $W(p, K, J)$ and $W(p, J)$ is a positive cone, i.e., if w_1, w_2 are in $W(p, K, J)[or \; W(p, J)]$, then $aw_1 + bw_2$ is in $W(p, K, J)[W(p, J)]$ for any positive constants a, b.*

Proof. It suffices to show that $W(p, K, J)$ is closed under addition. Assume (3.53) holds with $w = w_i, i = 1, 2$. Then

$$\int_J |y'|^p (w_1 + w_2) \; \leq \; K^p \left\{ \left(\int_J |y|^p w_1 \int_J |y''|^p w_1 \right)^{1/2} \right.$$

$$+ \left(\int_J |y|^p w_2 \int_J |y''|^p w_2 \right)^{1/2} \Bigg\}$$

$$\leq K^p \left(\int_J |y|^p (w_1 + w_2) \right)^{1/2} \left(\int_J |y''|^p (w_1 + w_2) \right)^{1/2},$$

where we used the Schwarz inequality for sums in the last step. □

3.3 Monotone Weight Functions

The next result plays a fundamental role in much of what follows.

Theorem 3.7 *Let $J = (a, b)$, $-\infty \leq a < b \leq \infty$, let p', q' be a conjugate pair $1/p' + 1/q' = 1$, $1 < p', q' < \infty$. Suppose f, g, h are nonnegative functions on J. If there is a constant K such that*

$$\int_c^b g \leq K \left(\int_c^b f \right)^{1/p'} \left(\int_c^b h \right)^{1/q'} \tag{3.55}$$

for every c in (a, b) — in particular, f, g, h are integrable on (c, b) for each c in (a, b) — then for any nondecreasing weight function w we have

$$\int_c^b gw \leq K \left(\int_c^b fw \right)^{1/p'} \left(\int_c^b hw \right)^{1/q'} \tag{3.56}$$

for all $c \in [a, b)$ provided the right-hand side is finite.

Proof. This proof is contained in the proof of the next lemma. See also the remarks following Lemma 3.5. □

The next result extends (3.56) to the case when the weight function w appearing in the three integrals in (3.56) can be different in the different integrals. We will need this extension in Section 4.

Lemma 3.5 *Let $J = (a, b)$, $-\infty \leq a < b \leq \infty$, and let p', q' satisfy $1 < p', q' < \infty$, $1/p' + 1/q' = 1$. Suppose s is a nonnegative function such that $s^{p'}$ and $s^{-q'}$ are integrable on $[0, T]$ for all $T > 0$ and define*

$$u(t) = \int_0^t s^{p'}, \quad v(t) = \int_0^t s^{pq'}, \quad t \geq 0. \tag{3.57}$$

Assume f, g, h are nonnegative functions such that (3.55) holds for all c in (a, b) with the same constant K (in particular f, g, h are integrable on (c, b) for each c in (a, b)). Then for any nondecreasing weight function w we have

$$\int_c^b g(t)w(t)dt \leq K \left(\int_c^b f(t)u(w(t))dt \right)^{1/p'} \left(\int_c^b h(t)v(w(t))dt \right)^{1/q'} \qquad (3.58)$$

for all c in $[a, b)$ provided the two integrals on the right of (3.58) are finite when $c = a$. (The finiteness of $\int_a^b gw$ is part of the conclusion.)

Before giving the proof of Lemma 3.5 we make some remarks. Lemma 3.5 states that if (3.55) holds for a given triple of functions f, g, h over all subintervals (c, b) of (a, b) with a uniform constant K, i.e., K does not depend on c, then an inequality of the same form, i.e., (3.58) holds with the same constant K when nondecreasing weight functions $w, u(w), v(w)$ are included in the integrands. Note that (3.58) holds with the same constant K for **all** nondecreasing weights w. Also, notice that the hypothesis (3.55) is assumed to hold only for $c > a$. In particular, it may happen that one or both of f, h are not integrable on (a, b) but (3.58) may hold even for $c = a$ when w is small enough near a so that $fu(w)$ and $hv(w)$ are integrable on (a, b).

An analogous result holds for inequalities of sum form: If

$$\int_c^b g \leq K \left(\int_c^b f + \int_c^b h \right)$$

for all c in (a, b), then

$$\int_c^b gw \leq K \left(\int_c^b fw + \int_c^b hw \right)$$

holds for all c in (a, b) and all nondecreasing weight functions w. Note that we use the same weight w in all three integrals here. The proof is similar and hence omitted.

To illustrate (3.58) we mention a couple of simple examples.

(i) When $s(t) = 1$, $u(w(t)) = w(t)$, $v(w(t)) = w(t)$, then (3.58) reduces to (3.56).

(ii) When $s(t) = (1 + t)^{1/2}$, $u(w) = w + w^2/2$, $v(w) = \ln(1 + w)$.

(iii) When $s(t) = (1 + t^2)^{1/2}$, $u(w) = w + w^3/3$, $v(t) = \tan^{-1} w$.

A variety of inequalities (3.58) can be generated by making different choices for s. Other choices of s will appear in some of our results below.

The reflection that maps (a, b) onto (b, a) yields a result dual to Lemma 3.5: If (3.55) with (c, b) replaced by (a, c) holds for each c in (a, b) with a uniform constant K, then (3.56) and (3.58) hold with (c, b) replaced by (a, c) for each c in $(a, b]$ and any nondecreasing weight function w provided the two integrals on the right are finite. Similarly, there is a dual result for the sum form inequality mentioned above.

Repeated applications of Lemma 3.5 yield the following lemma.

Lemma 3.6 *Assume the hypothesis of Lemma 3.5. Suppose s_j are nonnegative functions such that $s_j^{p'}$ and $s_j^{-q'}$ are integrable on $[0, T)$ for each $T > 0$ and let u_j, v_j be defined by (3.57) with $s = s_j$, $j = 1, \ldots, m$. Then (3.55) implies that for all $c \in [a, b)$*

$$\int_c^b g(t) w_1(t), \ldots, w_m(t) dt \leq K \left(\int_c^b f(t) u_1(w_k(t)) \ldots u_m(w_m(t)) dt \right)^{1/p'}$$
$$\left(\int_c^b h(t) v_1(w_1(t)) \ldots v_m(w_m(t)) dt \right)^{1/q'} \tag{3.59}$$

irrespective of the choice of the nondecreasing weight functions $w_1, \ldots, w_m$. (As before (3.59) holds for all c in (a, b) and with $c = a$ if the two integrals on the right are finite.)

Proof. Lemma 3.6 follows from repeated applications of Lemma 3.5. □

Proof. (of Lemma 3.5) It is sufficient to establish the case $a > -\infty$. The case $a = -\infty$ then follows by letting $a \to -\infty$. Next we note that it is enough to prove inequality (3.58) when $c = a$. One of the difficulties is that the integrability of the left-hand side of (3.55) is not known in advance. To overcome this we make use of a truncation of the weight function

$$w(t; t_1, t_2) = \begin{cases} 0, & a < t < t_1, \\ w(t), & t_1 \leq t \leq t_2, \\ w(t_2), & t_2 < t. \end{cases}$$

Inequality (3.58) will be proved for $w(t; t_1, t_2)$ in place of w. Letting $t_1 \to a$ and $t_2 \to b$ yields (3.58). Thus we may assume that w has the following properties: $w = 0$ in a neighborhood (a, t_1) of a so that gw is integrable on (a, t_1) and w is a constant outside a compact interval so that $z = w(t_2) = \max w < \infty$, implying that gw, $fu(w)$, and $hv(w)$ are all integrable on (a, b).

Define the function $A(t, y)$ on the strip $S = \{(t, y) : a < t < b, 0 \leq y \leq z\}$ by

$$A(t, y) = \begin{cases} 1, & \text{if } y \leq w(t), \\ 0, & \text{if } y > w(t) = w(t; t_1, t_2). \end{cases}$$

Clearly, we have

$$\int_0^z A(t, y) dy = w(t)$$

for all $t \in (a, b)$. Hence

$$\int_a^b g(t) w(t) dt = \int_a^b g(t) \left\{ \int_0^z A(t, y) dy \right\} dt$$

and, changing the order of integration, we get

$$\int_a^b g(t)w(t)dt = \int_0^z \left\{ \int_{k(y)}^b g(t)dt \right\} dy, \tag{3.60}$$

where $k(y) = \min_t \{A(t,y) = 1\}$.

Using hypothesis (3.55) with $c = k(y)$ and (3.60) we get

$$\int_a^b g(t)w(t)dt \leq K \int_0^z \left\{ \left\{ \int_{k(y)}^b f(t)dt \right\}^{1/p'} \left\{ \int_{k(y)}^b h(t)dt \right\}^{1/q'} \right\} dy$$

$$\leq K \left\{ \int_0^z s^{p'}(y) \int_{k(y)}^b f(t)dtdy \right\}^{1/p'}$$

$$\left\{ \int_0^z s^{-q'}(y) \int_{k(y)}^b h(t)dtdy \right\}^{1/q'}. \tag{3.61}$$

Hölder's inequality was used in the second step. With the order of integration changed, the first integral on the right of (3.61) becomes

$$\int_S \int s^{p'}(y)f(t)A(t,y)dtdy = \int_a^b \left(\int_0^{w(t)} s^{p'}(y)dy \right) f(t)dt$$

$$= \int_a^b f(t)u(w(t))dt. \tag{3.62}$$

By rewriting the second integral on the right of (3.61) similarly we obtain (3.58). This completes the proof of Lemma 3.5. □

Theorem 3.8 *Let $J = R = (-\infty, \infty)$ or $J = R^+ = (0, \infty)$, and n, k be integers satisfying $1 \leq k < n$. Let $1 \leq p, r < \infty$ and choose q by*

$$nq^{-1} = (n-k)p^{-1} + kr^{-1}. \tag{3.63}$$

Let α, β be given by

$$\alpha = (n - k - r^{-1} + q^{-1})/(n - r^{-1} + p^{-1}), \quad \beta = 1 - 2. \tag{3.64}$$

Let $s(t)$ be a nonnegative function defined for $t \geq 0$ such that

$$s^{p/(\alpha q)}, s^{-r/(\beta q)} \in L^1(0,T), \text{ all } T > 0. \tag{3.65}$$

Define u, v by (3.57) with $p' = p/(\alpha q)$, $q' = r/(\beta q)$. Then for any nondecreasing weight function w, if

$$\int_J |y(t)|^p u(w(t)) < \infty \text{ and } \int_J |y^{(n)}(t)|^r v(w(t))dt < \infty, \tag{3.66}$$

it follows that

$$\left(\int_J |y^{(k)}(t)|^q w(t) dt \right)^{1/q} \le K \left(\int_J |y(t)|^p u(w(t)) dt \right)^{\alpha/p} \left(\int_J |y^{(n)}(t)|^r v^r(w(t)) dt \right)^{\beta/r}, \quad (3.67)$$

where $K = K(n,k,p,q,r,R^+)$. *More generally,*

$$\left(\int_J |y^{(k)}(t)|^q w_1(t) \ldots w_m(t) dt \right)^{1/q} \le K \left(\int_J |y(t)|^p u_1(w_1(t)) \ldots u_m(w_m(t)) dt \right)^{\alpha/p}$$

$$\left(\int_J |y^{(n)}(t)|^r v_1(w_1(t)) \ldots v_m(w_m(t)) dt \right)^{\beta/r}, \quad (3.68)$$

where w_i, u_i, v_i, $i = 1, \ldots, m$ *are as in Lemma 3.6.*

Proof. Theorem 3.8 follows from Lemmas 3.5, 3.6, and Theorem 1.4 of Chapter 1, Section 3. Clearly, (3.67) and (3.68) hold if w is identically zero. Since w is nondecreasing we may assume that $w(t) > 0$, for $t \ge a$. Let $c > a$. Then by (3.57), $u(w(c)) > 0$ and $v(w(c)) > 0$. Since $u(w)$ and $v(w)$ are nondecreasing, (3.66) implies that

$$\int_c^\infty |y^p| < \infty \text{ and } \int_c^\infty |y^{(n)}|^r < \infty.$$

Hence, by Theorem 1.4 of Chapter 1, Section 3, (3.67) holds with $w = u(w) = v(w) = 1$ and $K = K(n,k,p,q,r,R^+)$ on each interval $J = J_c = (c,\infty)$, $c > a$. Now apply Lemma 3.5 with $(a,b) = (a,\infty)$, $f = |y|^p$, $g = |y^{(k)}|^q$, $h = |y^{(n)}|^r$, $p' = p/(\alpha q)$, $q' = r/\beta q$ and K replaced by K^2. This gives (3.67). Note that p' and q' are conjugate by (3.63). This is where we need equality in (3.63) — see Remarks below. Similarly (3.68) follows from Lemma 3.6 and Theorem 1.4 of 1.3. □

An important special case of Theorem 3.8 is listed as the following corollary.

Corollary 3.2 *Let* $J = R$ *or* R^+, *let the integers* n,k *satisfy* $1 \le k < n$, *and let* $1 \le p$, $r < \infty$. *Let* q *be determined by (3.63),* α, β *by (3.65). If*

$$\int_J |y|^p w < \infty \text{ and } \int_J |y^{(n)}|^r w < \infty$$

for some nondecreasing weight function w *then*

$$\left(\int_J |y^{(k)}|^q w \right)^{1/q} \le K \left(\int_J |y|^p w \right)^{\alpha/p} \left(\int_J |y^{(n)}|^r w \right)^{\beta/r}, \quad (3.69)$$

where $K = K(n,k,p,q,r,R^+)$.

Proof. This is the special case $s(t) = 1$ of Theorem 3.7, since $s(t) = 1$ implies $u(w) = w = v(w)$. □

We now make a number of remarks about Theorem 3.8.

Remark 3.5 Write (3.69) as

$$\|y^{(k)}\|_{q,w} \leq K \, \|y\|_{p,w} \, \|y^{(n)}\|_{r,w}. \tag{3.70}$$

In comparing (3.70) with the special case $w = 1$, i.e., inequality (1.25) of 1.3, we note that (1.25) was established for all p, q, and r satisfying the inequality

$$nq^{-1} \leq (n-k)p^{-1} + kr^{-1}, \tag{3.71}$$

whereas in Theorem 3.7 we assumed equality in (3.71). Does Corollary 3.2 hold when (3.63) is replaced by (3.71)? That the answer is no can be seen from the following considerations. Assume strict inequality holds in (3.71). Let $J = R^+$, $w(t) = t$. Choose a C_0^∞ function $g \not\equiv 0$ with $g^{(n)} \not\equiv 0$ and consider the quotient

$$Q(g) \; = \; \left(\int_J |g^{(k)}(t)|^q t \, dt \right)^{1/q} \left(\int_J |g(t)|^p t \, dt \right)^{-\alpha/p}$$
$$\left(\int_J |g^{(n)}(t)|^r t \, dt \right)^{-\beta/r}.$$

Let $g_a(t) = g(at)$, $a > 0$. Then $Q(g_a) = a^c Q(g)$, where $c = \alpha p^{-1} + \beta r^{-1} - q^{-1} > 0$. Hence the quotients $Q(g_a)$, $a > 0$ are not bounded above and hence there is no finite constant K for which (3.69) holds in this case.

Corollary 3.3 *Let $J = (-\infty, a)$, $-\infty < a \leq \infty$. Let $p, q, r, n, k, \alpha, \beta$ be as in Theorem 3.8. If w is a weight function which is **nonincreasing**, then (3.69) holds with the same K provided the two integrals on the right are finite.*

Proof. This follows from Corollary 3.2 and the change of variable $t \to -t$. $\square$

Corollary 3.4 *Let $J = R = (-\infty, \infty)$ and let $p, q, r, n, k, \alpha, \beta$ be as in Theorem 3.8. If w is a weight function which is either nondecreasing or nonincreasing, then (3.69) holds provided the two integrals on the right are finite.*

Proof. This follows from Corollaries 3.2 and 3.3. $\square$

Remark 3.6 Note that in Theorem 3.8 and its Corollaries, the weight functions w were not assumed continuous.

Corollary 3.5 *Let $J = (a, \infty)$, $-\infty \leq a < \infty$, let $p, q, r, \alpha, \beta, n, k$ be as in Theorem 3.8. If w is a nondecreasing weight function then*

$$\left(\int_J |y^{(k)} w|^q \right)^{1/q} \leq AK \left(\int_J |yw|^p \right)^{\alpha/p} \left(\int_J |y^{(n)} w|^r \right)^{\beta/r}, \tag{3.72}$$

where

$$K = K(n, k, p, q, r, R^+), \quad A = q^{1/q} p^{-\alpha/p} r^{-\beta/r}, \tag{3.73}$$

provided the two integrals on the right are finite.

Proof. Take $s(t) = t^c$, $c = \alpha(p-q)/p = \beta(q-r)/r$. Then $u(t) = q t^{p/q}/p$, $v(t) = q t^{r/q}/r$. Now use (3.66) with w replaced by w^q. $\square$

In Theorem 3.8 the interval J is unbounded. The next result applies to bounded intervals J.

Theorem 3.9 *Let $J = (a, b)$, $-\infty \le a < b < \infty$. Let $p, q, r, n, k, \alpha, \beta$ satisfy the conditions of Theorem 3.8. If w is any nondecreasing weight function, then there is some constant K such that (3.67) holds for all $y \in W_{p,r}^n(J)$ satisfying*

$$y^{(m)}(t_m) = 0 \text{ for some } t_m \text{ in } J, \ m = 1, \ldots, n-1. \tag{3.74}$$

Proof. This follows from Corollary 1.1 of 1.3 and Lemma 3.5. $\square$

Definition. Let $K = K(n, k, p, q, r, J, w)$ denote the smallest constant in inequality (3.69). Let $K(n, k, p, q, r, J, 1) = K(n, k, p, q, r, J)$ as in Chapters 1 and 2.

Next we investigate the relationship between the best constants for general weights w and the classical ones when $w = 1$ on the whole line or half line.

Theorem 3.10 *Let $p, q, r, n, k, \alpha, \beta$ be as in Theorem 3.8. Then the inequalities*

$$K(n, k, p, q, r, R) \le K(n, k, p, q, r, J, w) \le K(n, k, p, q, r, R^+) \tag{3.75}$$

hold for any nondecreasing weight function $w \ne 0$ and any interval $J = (a, \infty)$, $-\infty \le a < \infty$.

Furthermore, if the support of w is contained in

$$J = [a, \infty), \ -\infty < a \text{ and } \lim_{t \to a^+} w(t) > 0, \tag{3.76}$$

then

$$K(n, k, p, q, r, J, w) = K(n, k, p, q, r, R^+). \tag{3.77}$$

Proof. The right inequality in (3.75) follows from Corollary 3.2, i.e., (3.69). To establish the other half let $K = K(n, k, p, q, r, R)$. Choose a point t_0 in the interior of the support of w at which w is continuous. Such a point exists since w is nondecreasing by hypothesis and hence its set of discontinuities is at most countable. By considering the translation $t \to t - t_0$ we may assume, without loss of generality, that $t_0 = 0$. Thus we have

$$\lim_{\epsilon \to 0}[\inf w(t)/ \sup w(t)] = 1, \tag{3.78}$$

where both the inf and sup are taken over the interval $(-\epsilon, \epsilon)$, $\epsilon > 0$. Let $\delta > 0$. By Lemma 1.4 of 1.3 and the definition of K there exists a function $g \in W_{p,r}^n(J)$ with compact support such that

$$Q(g) = \|g^{(k)}\|_q \, \|g\|_p^{-\alpha} \, \|g^{(n)}\|_r^{-\beta} > K - \delta. \tag{3.79}$$

Noting that $\alpha/p + \beta/r = 1/q$ we see that the left side of (3.79) is invariant under the change of variable $t \to \lambda t$. In other words, the left-hand side of (3.79) is unchanged if g is replaced by g_λ, where $g_\lambda(t) = g(\lambda t)$, $\lambda > 0$. Thus we can assume that the support of g is contained in the interval $(-\epsilon, \epsilon)$. Hence

$$\left(\int_J |g^{(k)}|^q w\right)^{1/q} \left(\int_J |g|^p w\right)^{-\alpha/p} \left(\int_J |g^{(n)}|^r\right)^{-\beta/r}$$

$$= \left(\int_{-\epsilon}^{\epsilon} |g^{(k)}|^q w\right)^{1/q} \left(\int_{-\epsilon}^{\epsilon} |g|^p w\right)^{-\alpha/p} \left(\int_{-\epsilon}^{\epsilon} |g|^p w\right)^{-\beta/r}$$

$$\geq \left(\inf_{-\epsilon < t < \epsilon} w(t)/ \sup_{-\epsilon < t < \epsilon} w(t)\right)^{1/q} \|g^{(k)}\|_q \, \|g\|_p^{-\alpha/p} \|g\|_r^{-\beta/r}$$

$$> \left(\inf_{-\epsilon < t < \epsilon} w(t)/ \sup_{-\epsilon < t < \epsilon} w(t)\right)^{1/q} (K - \delta). \tag{3.80}$$

By (3.78), (3.79) is greater than $K - \delta$ for ϵ small enough. Now let $\delta \to 0$. This completes the proof of (3.75).

The proof of the furthermore part of Theorem 3.10 is established by showing that

$$K(n, k, p, q, , r, J, w) \geq K(n, k, p, q, r, R^+)$$

using the same technique and exploiting $W_{p,r}^n(J)$ functions having support in a small right neighborhood of a. $\square$

3.4 Positive Weight Functions

In this section we consider weight functions which are not necessarily monotone but are strictly positive.

Theorem 3.11 *Let $1 \leq p < \infty$. Suppose b, c, d are real numbers satisfying the conditions*

$$(i) \; 2c = b + d, \qquad (ii) \; b > -1 - p, \qquad (iii) \; c > -1. \tag{3.81}$$

Then there is a constant $K = K(b, c, d, p)$ such that for all $a \geq 0$ we have

$$\left(\int_a^\infty t^c |y'(t)|^p dt \right)^2 \leq K \int_a^\infty t^b |y(t)|^p dt \int_a^\infty t^c |y''(t)|^p dt. \qquad (3.82)$$

for all admissible functions y.

Proof. In the classical case when $b = c = d = 0$ the inequality (3.82) for $a > 0$ follows from the case $a = 0$ by simply considering the translation $t \to t + a$. However, under the conditions here the case $a > 0$ does not seem to follow easily from the case $a = 0$.

The proof of Theorem 3.11 will be established with the help of several lemmas. Below the symbol K will denote a constant. However, it need not be the same constant at each occurrence. $\square$

Lemma 3.7 *There is a constant K such that for any interval I of unit length we have*

$$\int_I |y'|^p, \ \|y'\|_\infty^p \leq K \left(\int_I |y|^p + \int_I |y''|^p \right). \qquad (3.83)$$

Proof. This follows from Lemma 1.2 of 1.2. $\square$

Lemma 3.8 *Let K be as in (3.83). Then for any interval I of length $L < \infty$ we have*

$$\int_I |y'|^p \leq K \left(L^{-p} \int_I |y|^p + L^p \int_I |y''|^p \right). \qquad (3.84)$$

Proof. Without loss of generality we may take the left end point of I to be 0. Let $g(t) = y(Lt)$, $0 \leq t \leq 1$. By (3.83)

$$\int_0^1 L^p |y'(Lt)|^p \, dt \leq K \left(\int_0^1 |y(Lt)|^p \, dt + \int_0^1 L^{2p} |y''(Lt)|^p \, dt \right).$$

The change of variable $x = Lt$ gives (3.84). $\square$

Lemma 3.9 *Let b, c, d denote real numbers satisfying*

$$2c = b + d, \qquad (3.85)$$

$$b > c - p. \qquad (3.86)$$

Let $I = [\alpha, \beta]$ be such that $\alpha \geq 1$, $\lambda = \beta - \alpha = \alpha^h < \alpha$ where $h = (c - b)/p < 1$. Then

$$\int_I t^c |f'(t)|^p dt \ \leq \ K 2^{|b| + |b - c|} \left(\int_I t^b |f(t)|^p dt + \int_I t^d |f''(t)|^p dt \right). \qquad (3.87)$$

Proof. Note that for each real number a

$$t^a \leq 2^{|a|} \min_I t^a, \quad \max_I t^a \leq 2^{|a|} t^a. \tag{3.88}$$

Now we have, using (3.86) and Lemma 3.8,

$$
\begin{aligned}
\int_I t^c |f'(t)|^p dt &\leq 2^{|c|} \int_I \min t^c |f'(t)|^p dt \\
&\leq K 2^{|c|} \left(\int_I t^c \lambda^{-p} |f(t)|^p dt + \int_I t^c \lambda^p |f''(t)|^p dt \right) \\
&= K 2^{|c|} \left(\int_I t^c \alpha^{b-c} |f(t)|^p dt + \int_I t^c \alpha^{d-c} |f''(t)|^p dt \right) \\
&\leq K 2^{|c|+|b-c|} \left(\int_I t^b |f(t)|^p dt + \int_I t^d |f''(t)|^p dt \right). \qquad \square
\end{aligned}
$$

Lemma 3.10 *Let* $a_1 \geq 1$, $a_{n+1} = a_n + a_n^h$, $n = 1, 2, 3, \ldots$ *for any real number* h. *Then* $a_n \to \infty$.

Proof. For $h \geq 0$, $a_{n+1} - a_n \geq 1$, hence $a_n \to \infty$. For $h < 0$, note first that the sequence a_n is increasing. Suppose $a_n \to b < \infty$. Then $b - a_n < b^h < a_n^h$ for n large enough. This implies that $a_{n+1} > b$, a contradiction. $\square$

Lemma 3.11 *Let* b, c, d *satisfy (3.85) and (3.86). Then there is a* $K > 0$ *such that for each* $a \geq 1$ *we have*

$$\int_a^\infty t^c |f'(t)|^p dt \leq K \left(\int_a^\infty |t^b| f(t)^p dt + \int_a^\infty t^d |f''(t)|^p dt \right). \tag{3.89}$$

Proof. Let $a_1 = a$ and define a_n as in Lemma 3.10. Apply (3.87) to each interval (a_n, n_{n+1}) and sum over these intervals. $\square$

Lemma 3.12 *Let* b, c, d *satisfy (3.85),(3.86) and*

$$c > -1. \tag{3.90}$$

Then there exists a constant K *such that for any* a *in [0,1] we have*

$$\int_0^1 t^c |f'(t)|^p dt \leq K \left(\int_a^\infty t^b |f(t)|^p dt + \int_a^\infty t^d |f''(t)|^p dt \right). \tag{3.91}$$

Proof. By (3.83) and the fact that t^b, t^d are bounded away from zero on [1,2] we have

$$|f'(1)|^p \leq K \left(\int_1^2 t^b |f(t)|^p dt + \int_1^2 t^d |f''(t)|^p dt \right). \tag{3.92}$$

For each t in $[a, 1]$ we have, with $p^{-1} + q^{-1} = 1$

$$
\begin{aligned}
|f'(t)| &\leq |f'(1)| + \int_t^1 |f''(x)| dx \leq |f'(1)| + \left\{ \int_t^1 x^{-dq/p} dx \right\}^{1/2} \\
&\quad \left\{ \int_t^1 x^d |f''(x)|^p dx \right\}^{1/p} \\
&\leq |f'(1)| + K\{1 + t^{(1/q - (d/p))} + |\log(t)|^{1/q}\} \left\{ \int_a^1 t^d |f''(t)|^p dt \right\}^{1/p}.
\end{aligned}
$$

Hence

$$
|f'(t)|^p \leq K \left\{ |f'(1)|^p + (1 + t^{(p/q)-d} + |\log(t)|^{p/q}) \int_a^1 t^d |f''(t)|^p \right\} dt.
$$

Note that $c + (p/q) - d > (p/q) - p = p(q^{-1} - 1) = -1$. Therefore

$$
\begin{aligned}
\int_a^1 t^d |f'(t)|^p dt &\leq K \left\{ |f'(1)|^p \int_a^1 t^c dt + \int_a^1 t^d |f''(t)|^p dt \int_a^1 (t^{c+(p/q)-d} + t^c |\log(t)|^{p/q}) dt \right\} \\
&\leq K_1 \left\{ |f'(1)|^p + \int_a^1 t^d |f''(t)|^p dt \right\}
\end{aligned} \tag{3.93}
$$

since

$$
\int_a^1 t^d dt \leq \int_0^1 t^c dt < \infty, \quad \int_a^1 t^{c+(p/q)-d} dt \leq \int_0^1 t^{c+(p/q)d} < \infty
$$

and

$$
\int_a^1 t^c |\log t|^{p/q} dt \leq \int_0^1 t^{c-\epsilon} t^\epsilon |\log t|^{p/q} dt < \infty.
$$

The conclusion follows from (3.92) and (3.93). $\square$

Proof. (of Theorem 3.11). From Lemmas 3.11 and 3.12 we may conclude that there is a constant K such that

$$
\int_a^\infty t^c |f'(t)|^p dt \leq K \left\{ \int_a^\infty t^b |f(t)|^p dt + \int_a^\infty t^d |f''(t)|^p dt \right\} \tag{3.94}
$$

for all $a \geq 0$, provided that (3.85), (3.86), and (3.90) hold.

Now replacing t by λt in (3.94) and then minimizing over $\lambda > 0$ we get the product form of (3.94), namely (3.82).

This proves Theorem 3.11 under the stronger condition (3.86) instead of (ii).

Since the constant K in (3.82) is independent of a when $a \geq 0$, we can apply Lemma 3.5 of Section 3 to obtain

$$
\left\{ \int_a^\infty t^{c'} |f'(t)|^p w(t) dt \right\}^2 \leq K \int_a^\infty t^{b'} |f(t)|^p u(w(t)) dt \int_a^\infty t^{d'} |f''(t)|^p v(w(t)) dt \tag{3.95}
$$

for any b', c', d' satisfying (3.85), (3.86), and (3.90). Furthermore, (3.95) holds with the same constant K as in (3.72) when $c = c'$, $b = b'$, $d = d'$ and for any non-negative non-decreasing (locally integrable) function w.

So far we have established Theorem 3.11, i.e., inequality (3.82), only under the conditions (3.85), (3.86), (3.90). To see that (3.94) still holds when (3.86) is replaced by (ii), we use (3.95).

Let b, c, d satisfy (i), (ii), (iii) of Theorem 3.11, i.e., $2c = b + d$, $b > -1 - p$, $c > -1$. If $b > c - p$, we are done by (3.72). Suppose $b \leq c - p$. Choose c' so that $c' - p < b$ and $-1 < c' < c$. Now choose b' so that $c' - p < b' < b$ and

$$b - b' < 2(c - c'). \tag{3.96}$$

Let $d' = 2c' - b'$. Then (3.72) holds with b, d, c replaced by b', c', d', respectively. Let $w(t) = t^{c-c'}$ and $s(x) = x^\alpha$. by (3.57) with $p' = q' = 2$ we have

$$u(w) = (2\alpha + 1)^{-1} w^{2\alpha+1} \text{ if } 2\alpha > -1$$

and

$$v(w) = (-2\alpha + 1)^{-1} w^{-2\alpha+1} \text{ if } -2\alpha > -1.$$

Thus (3.95) will imply (3.94) if α satisfies

$$b' + (c - c')(2\alpha + 1) = b \tag{3.97}$$

and

$$d' + (c - c')(-2\alpha + 1) = d \tag{3.98}$$

and

$$-\frac{1}{2} < \alpha < \frac{1}{2}. \tag{3.99}$$

Choose α by (3.97). Then (3.88) is satisfied since these two equations are compatible and (3.99) follows from (3.96). This completes the proof of Theorem 3.11. □

Remark 3.7 Having established (3.94) under conditions (i), (ii), (iii) of Theorem 3.11, it follows from Theorem 3.13 that inequality (3.95) holds for any b', c', d' satisfying $c' > -1$, $2c' = b' + d'$, $b' > -1 - p$, with the same constant $K = K(b', c', d', p)$, for any non-negative non-decreasing weight function w.

Theorem 3.12 *Let p, b, c, d satisfy the conditions of Theorem 3.11. Assume (3.82) holds with the same constant K for all $a \geq 0$. Let u, v be given by (3.57). Then the inequality*

$$\left(\int_a^\infty t^c |y'(t)|^p w(t) dt \right)^2 \leq K \int_a^\infty t^b |y(t)|^p u(w(t)) dt \int_a^\infty t^d |y''(t)|^p v(w(t)) dt \tag{3.100}$$

holds for all $a \geq 0$ and all nondecreasing weight functions w with the same constant K.

Proof. This follows from Theorem 3.11 and Lemma 3.5 of Section 3. □

Theorem 3.13 *Let* $1 \leq p < \infty$, $c > -1$, *let* n, k *be integers with* $1 \leq k < n$. *Let* u, v *be defined by (3.57) with* $p' = \alpha^{-1}$, $q' = \beta^{-1}$, *where* α, β *are given by* $\alpha = (n - k)/n$, $\beta = k/n$. *Then there exists a positive constant* K *such that the inequality*

$$\int_a^\infty t^c |y^{(k)}(t)|^p dt \leq K \left(\int_a^\infty t^c |y(t)|^p dt \right)^\alpha \left(\int_a^t t^c |y^{(n)}(t)|^p dt \right)^\beta \tag{3.101}$$

holds for all $a \geq 0$. *Moreover, for the same constant* K *we have that the inequality*

$$\int_a^\infty t^c |y^{(k)}(t)|^p w(t) dt$$

$$\leq K \left(\int_a^\infty t^c |y(t)|^p u(w(t)) dt \right)^\alpha \left(\int_a^\infty t^c |y^{(n)}(t)|^p v(w(t)) dt \right)^\beta \tag{3.102}$$

holds for all $a \geq 0$ *and all nondecreasing weight functions* w. *In particular, (3.102) holds with* $u(w(t)) = w(t) = v(w(t))$.

Proof. Inequality (3.101) follows from (3.82) by induction. (Note that K in (3.101) is not the same, in general, as the K in (3.82).) Lemma 3.8 and (3.101) yield (3.102). The last sentence is the special case when $s(t) = 1$, so that $u(t) = t = v(t)$, in (3.57). □

Theorem 3.14 *Let* $1 \leq p < \infty$, $b \geq 0$, $c \geq 0$, *and let* n, k *be integers with* $1 \leq k < n$. *Let* $\alpha = (n - k)/n$, $\beta = k/n$. *Then there is a positive constant* K *such that*

$$\int_a^\infty t^{kb+c} |y^{(k)}(t)|^p dt \leq K \left(\int_a^\infty t^c |y(t)|^p dt \right)^\alpha \left(\int_a^\infty t^{nb+c} |y^{(n)}(t)|^p dt \right)^\beta, \tag{3.103}$$

for all $a \geq 0$ *(and all admissible* y).

Proof. In (3.102) take $c = -1/2$, $w(t) = t^\lambda$. Take $s(t) = t^\gamma$ and determine u and v by (3.3). Then γ and λ can be so chosen that (3.102) becomes (3.103). □

3.5 Weights with Zeros

In Section 2 we defined $W(p, K, J)$ to be the class of all weight functions w for which the inequality

$$\left(\int_J |y'|^p w \right)^{1/p} \leq K \left(\int_J |y|^p w \right)^{1/(2p)} \left(\int_J |y''|^p w \right)^{1/2\,p} \tag{3.104}$$

holds for all admissible functions y and $W(p, J)$ to be the union of $W(p, K, J)$ for all $K > 0$. In Sections 3 and 4 we found sufficient conditions for w to be in $W(p, J)$. These conditions are independent of p for $1 \leq p < \infty$. Here we show that $W(p, J)$ does depend on p. But our main concern in this section is the study of weight functions which may have one or more isolated zeros on J.

We start with an observation.

Theorem 3.15 *If $w \in W(p, J)$, $1 \leq p < \infty$, $J = R$ or R^+, then for any integer $n \geq 2$ and any integer k with $1 \leq k < n$ there exists some finite constant K such that*

$$\left(\int_J |y^{(k)}|^p w \right) \leq K \left(\int_J |y|^p w \right)^\alpha \left(\int_J |y^{(n)}|^p w \right)^\beta, \qquad (3.105)$$

where $\alpha = (n-k)/n$, $\beta = k/n$. Here (3.105) holds for all admissible y with the same K but, of course, the K in (3.105) need not be the same as the K in (3.104).

Proof. Inequality (3.105) follows from (3.104) and induction on n. □

Since, in this section, our main concern is the study of the class of weight functions $W(p, J)$ we will consider only (3.104) in the remainder of this section.

Theorem 3.16 *The class $W = W(p, R^+)$ contains all positive linear combinations of functions of the form*

$$(t + a)^c b(t) w(t), \qquad (3.106)$$

where $c > -1$, $a \geq 0$, w is a nondecreasing weight function, and b is measurable with $0 < b_0 \leq b(t) \leq B < \infty$ for some constants b_0, B, and all $t \geq 0$.

Proof. By Theorem 3.15 it is enough to show that $(t + a)^c w(t)$ is in W. Thus, it is sufficient to prove that w is in W implies bw is in W. To that end, let $w \in W$. Then

$$
\begin{aligned}
\left(\int_J |y'|^p bw \right)^{2/p} & \leq B^{2/p} \left(\int_J |y'|^p w \right)^{2/p} \\
& \leq B^{2/p} K^2 \left(\int_J |y|^p w \right)^{1/p} \left(\int_J |y''|^p w \right)^{1/p} \\
& \leq K B^{2/p} \left(b_0^{-1} \int_J |y|^p bw \right)^{1/p} \left(b_0^{-1} \int_J |y''|^p bw \right)^{1/p} \\
& = K^2 (B/b_0)^{2/p} \left(\int_J |y|^p bw \right)^{1/p} \left(\int_J |y''|^p bw \right)^{1/p}.
\end{aligned}
$$

This completes the proof of Theorem 3.16. □

Note that this proof establishes the following result.

Corollary 3.6 *Let $J = (a, \infty)$, where $-\infty \leq a < \infty$. If $w \in W(p, K, J)$ then $bw \in W(p, (B/b_0)^{1/p}K, J)$ provided b is measurable and $0 < b_0 \leq b(t) < B < \infty$ for all $t \in J$.*

Theorem 3.11 of Section 4 shows that there are decreasing weight functions in $W(p, R^+)$. However, functions in $W(p, R^+)$ cannot decrease too rapidly. Let $w(x) \leq cx^{-1-\epsilon}$, $J = (1, \infty)$, and take $y(x) = x$. Then inequality (3.104) does not hold when $\epsilon > p$ since the first integral on the right is finite and the second one zero while the integral on the left of (3.104) is positive.

Corollary 3.7 *If the positive weight function u does not decrease too rapidly in the sense that*

$$\sup_{t \in J, s \geq 0} \frac{u(t)}{u(t+s)} = M < \infty, \tag{3.107}$$

then $u \in W(p, M^{1/p}K(p, R^+, 1), R^+)$ for $1 \leq p < \infty$.

Proof. Condition (3.107) implies that

$$u(t) = b(t)w(t)$$

for $t \in J$, where $w(t)$ is nondecreasing and $1 \leq b(t) \leq M < \infty$. Thus the conclusion of Corollary 3.7 follows from Corollary 3.6 and Theorem 3.7 of Section 3. □

Note that the case $M = 1$ of Corollary 3.7 is equivalent to u being nondecreasing and thus is contained in Theorem 3.8 of Section 3.

Using the theory of semigroups of linear operators, a different bound for K can be established.

Theorem 3.17 *Suppose the weight function u satisfies (3.107) and $J = (a, \infty)$ where $-\infty \leq a < \infty$. Then $u \in W(p, [2M^{1/p}(M^{1/p} + 1)]^{1/2}, J)$.*

Proof. Let $A = d/dx$ be the differentiation operator on the classical weighted Banach space $L^p(J, u)$ with the usual norm. It is well known that A is the infinitesimal generator of a (C_0) semigroup $\{T(t) : t \geq 0\}$ given by

$$T(t)f(x) = f(x + t), \qquad x \in J, \quad t \geq 0.$$

We claim that

$$\|T(t)\| \leq M^{1/p}, \qquad t \geq 0. \tag{3.108}$$

This follows from

$$\|T(t)f\|_p^p = \int_J |f(t+s)|^p u(s)ds = \int_{a+t}^{\infty} |f(u)|^p u(-t)dt$$

$$\leq \int_{a+t}^{\infty} |f(u)|^p M u(t)dt \leq M\|f\|_p^p. \qquad (3.109)$$

Now, (3.108) implies that

$$\|Af\|_p^2 \leq 2M^{1/p}(M^{1/p}+1)\|f\|_p\|A^2f\|_p. \qquad (3.110)$$

To see this, take the semigroup extension of Taylor's formula

$$T(t)f = f + tAf + \int_0^t (t-s)T(s)A^2f \, ds \qquad (3.111)$$

for $f \in \text{Dom}(A^2)$. Taking norms in (3.111) and solving for $\|Af\|$ we obtain

$$\|Af\| \leq t^{-1}(M^{1/p}+1)\|f\| + 2^{-1}tM^{1/p}\|A^2f\|.$$

Minimizing the right side over $t > 0$ gives (3.110) and completes the proof of the theorem. $\square$

If A generates a (C_0) group of isometries on a Banach space $(X, |\cdot|)$, then $|Af|^2 \leq 2|A^2f||f|$ holds for all $f \in \text{Dom}(A^2)$ by Ditzian [1983]. Now let A generate a (C_0) group $\{T(t) : -\infty < t < \infty\}$ on $(X, \|\cdot\|)$ satisfying $N = \sup_{t\in R}\|T(t)\| < \infty$. Then $|\cdot|$ defined by

$$|f| = \sup_{t\in R}\|T(t)f\|$$

is an equivalent norm on X with respect to which $\{T(t) : t \in R\}$ is a (C_0) group of isometries. Moreover, for $f \in \text{Dom}(A^2)$,

$$\|Af\|^2 \leq |Af|^2 \leq 2|A^2f||f| \leq 2N^2 \|A^2f\| \|f\|.$$

Thus for $a = -\infty$ we can improve the conclusion of Theorem 3.17 to read $u \in W(p, 2M^{1/(2p)}, R)$.

It is interesting to compare the bound $K = [2M^{1/p}(M^{1/p}+1)]^{1/2}$ of Theorem 3.17 with the bound $K = M^{1/p}K(p, R^+, 1)$ of Corollary 3.7. Since the exact values of $K(p, R^+, 1)$ are known only for $p = 1, 2$, and ∞, a precise comparison can be made only in these cases. In the other cases $(1 < p < \infty, p \neq 2)$, the upper bound $K(p, R^+, 1) \leq 2$ can be used. For $p = 2$, Corollary 3.7 gives a better constant K for all values of $M \geq 1$. Since $\lim_{p\to 2} K(p, R^+, 1) = \sqrt{2}$, Corollary 3.7 gives a better constant K for p near 2 and M bounded. When $p = 1$, $K(1, R^+, 1) = \sqrt{5/2}$ and so Corollary 3.7 gives a better constant for $M < 4$, whereas Theorem 3.17 yields a smaller K for $M > 4$. There is agreement when $M = 4$. Since $K(p, R^+, 1)$ is a continuous function of p (Section 9 of Chapter 2), we see that for p near 1, Corollary 3.7 gives a better constant for "small" values of $M \geq 1$ whereas Theorem 3.17 yields a smaller value of K for M large.

Theorems 3.16 and 3.17 do not allow a weight function w for which inequality (3.104) holds to have a zero at some point x_0 without being identically zero to the right or left of x_0. Here we consider weight functions w which have isolated zeros. It turns out that the validity of inequality (3.104) in such cases is a rather delicate matter which depends on the relationship between the order of such zeros and the value of p in (3.104).

Theorem 3.18 *Suppose w is a weight function, i.e., satisfies (3.54) on $J = (a, \infty)$, $-\infty \le a < \infty$. Then w is in $W(p, J)$ if the following conditions are satisfied: There exist constants c_1, c_2 such that if I is any compact subinterval of J and I_i, $i = 1, 2, 3$ denote the first, second, and third thirds of I, respectively, then*

$$(*) \qquad \int_I w / \int_{I_i} w \le c_1, \quad i = 1, 3,$$

$$(**) \qquad \int_I w^{-q/p} < \infty \text{ for } 1 < p < \infty, \quad p^{-1} + q^{-1} = 1,$$

$$\text{or } \int_I w^{-1} < \infty \text{ if } p = 1,$$

$$(***) \qquad \int_I w \left(\int_I w^{-q/p} \right)^{p/q} \le c_2 \, |I|^p \text{ for } 1 < p < \infty,$$

$$\text{or } \left(\int_I w \right) \sup_I w \le c_2 \, |I| \text{ if } p = 1,$$

where $|I|$ denotes the length of I.

Proof. The proof is based on a special case of Theorem 3.8 of Section 3 which we state here for the convenience of the reader as

Lemma 3.13 *If $w^{-q/p} \in L^1(I)$ for $1 < p < \infty$, or if $w^{-1} \in L^\infty(I)$ for $p = 1$, then*

$$\int_I w|y'|^p \le B \int_I w \int_I |y|^p w + A \int_I |y''|^p w \qquad (3.112)$$

for all admissible functions y, where

$$A = 2^{p-1} \int_I w \left(\int_I w^{-q/p} \right)^{p/q} \text{ if } 1 < p < \infty,$$

$$A = \left(\int_I w \right) \sup_I w^{-1} \text{ if } p = 1,$$

$$B = 2^{2(p-1)} |I|^{-p} \int_I w / \left(\min_{i=1,3} \int_{I_i} w \right).$$

Returning to the proof of Theorem 3.18, let $J = \sum_{i=1}^{\infty} I_i$, where I_i has length L and the interiors of the I_i's are disjoint. Using hypotheses $(*)$, $(***)$ in Theorem 3.18 we get

$$\int_J |y'|^p w \le c_1 2^{2(p-1)} L^{-p} \int_J |y|^p w + c_2 2^{p-1} L^p \int_J |y''|^p w.$$

Minimizing the right side of this inequality over all $L > 0$ gives

$$\int_J |y'|^p w \leq 2(c_1 2^{2(p-1)} c_2 2^{p-1})^{1/2} \left(\int_J |y|^p w \int_J |y''|^p w \right)^{1/2}.$$

This completes the proof of Theorem 3.18. $\square$

Theorem 3.19 *Let* $J = (a, \infty)$, $-\infty \leq a < \infty$. *Suppose w has an isolated zero at x_0 of (exact) order r and is C^r in a neighborhood of x_0. Then (3.104) does not hold, that is, $w \notin W(p, J)$ for $1 \leq p < r + 1$. On the other hand, there exist weight functions w having an isolated zero at x_0 of (exact) order r such that (3.104) holds, i.e., w is in $W(p, J)$, for $r + 1 < p$.*

Proof. For the first assertion it is enough to construct a sequence of functions $\{y_n\}$ such that $Q(y_n) \to \infty$ and $n \to \infty$, where

$$Q(y) = \left(\int_J |y'|^p w \right)^2 / \left(\int_J |y|^p w \int_J |y''|^p w \right).$$

We may assume without loss of generality that $J = (0, \infty)$. Let

$$y_n(x) = \begin{cases} x_0 - x & \text{if } 0 \leq x \leq x_0 - 1/n, \\ n^2(x - x_0)^3 + 2n(x - x_0)^2 & \text{if } x_0 - 1/n \leq x \leq x_0, \\ 0 & \text{if } x_0 \leq x. \end{cases}$$

Note that $y_n(x_0) = 0$, $y_n(x_0 - 1/n) = 1/n$, and

$$y_n'(x) = \begin{cases} -1 & \text{if } 0 \leq x \leq x_0 - 1/n, \\ 3n^2(x - x_0)^2 + 4n(x - x_0) & \text{if } x_0 - 1/n \leq x \leq x_0, \\ 0 & \text{if } x_0 \leq x, \end{cases}$$

$$y_n''(x) = \begin{cases} 0 & \text{if } 0 \leq x < x_0 - 1/n, \\ 6n^2(x - x_0) + 4n & \text{if } x_0 = 1/n < x < x_0, \\ 0 & \text{if } x_0 < x. \end{cases}$$

Also, we have

$$\int_J |y_n'(x)|^p w(x) dx \geq \int_0^{x_0 - 1/n} w(x) dx > 2^{-1} \int_0^{x_0} w(x) dx = K_1 > 0$$

for n large enough. Moreover, again for large n,

$$\begin{aligned} \int_J |y_n|^p w &\leq \int_0^{x_0 - 1/n} |x - x_0|^p w(x) dx + \int_{x_0 - 1/n}^{x_0} (2n^{-1})^p w(x) dx \\ &\leq \int_{1/n}^{x_0} t^p w(x_0 - t) dt + 1 \leq \int_0^{x_0} t^p w(x_0 - t) dt + 1 \\ &= K_0 > 0; \\ \int_J |y_n''|^p w &= \int_{x_0 - 1/n}^x [6n^2(x - x_0) + 4n]^p w(x) dx \\ &\leq 2^{p-1} \int_{x_0 - 1/n}^{x_0} [6^p n^{2p} x - x_0^p + 4^p n^p] w(x) dx = u(n, p). \end{aligned}$$

To complete the proof it suffices to show $u(n,p) \to 0$ as $n \to \infty$. Since w has a zero of order r at x_0 and is C^r in a neighborhood of x_0, it follows from Taylor's theorem that

$$w(x) = \frac{(x - x_0)^r}{r!} w^{(r)}(t_x) \le C|x - x_0|^r$$

for $x_0 - 1/n \le x \le x_0$. Thus

$$
\begin{aligned}
u(n,p) &\le 2^{p-1}C \int_{x_0-1/n}^{x_0} \{6^p n^{2p}|x - x_0|^p + 4^p n^p\}|x - x_0|^p dx \\
&= 2^{p-1}C \int_0^{1/n} (6^p n^{2p} t^{p+r} + 4^p n^p r^r) dt \\
&= 2^{p-1}C\{6^p n^{p-r-1}/(p+r+1) + 4^p n^{p-r-1}/(r+1)\} \to 0
\end{aligned}
$$

as $n \to \infty$ if $p < r + 1$.

To prove the final assertion of the Theorem, let

$$
w(x) = \begin{cases}
1 & \text{if } x \le x_0 - 1 \\
|x - x_0|^r & \text{if } x_0 - 1 \le x < x_0 + 1 \\
1 & \text{if } x > x_0 + 1.
\end{cases}
$$

Then w satisfies condition (*) of Theorem 3.18, and (**) and (***) are also satisfied when $p > r + 1$ as can be seen from a direct computation. $\square$

3.6   Notes and Problems

Section 1. This section is essentially taken from Kwong and Zettl [1981]. Amos and Everitt [1977] is an earlier paper on this subject. See also Evans and Zettl [1978].

Problem 1. Let u, v, w, n, k, and p satisfying the hypotheses of Theorem 3.4 be given. Let $J = (a, b)$ be a bounded interval. Then for any $\epsilon > 0$ there is a $K(\epsilon) > 0$ such that inequality (3.2) holds for all admissible functions y. What is the smallest value of $K(\epsilon)$? Redheffer [1963] showed that in the special case $k = 1$, $n = 2$, $u = v = w = 1$, and $p = 2$, the answer is

$$K(\epsilon) = 1/\epsilon + 12/(b - a)^2.$$

For $n > 2$ the best value of $K(\epsilon)$ does not seem to be known even for the case $p = 2$, $u = v = w = 1$. In the second order case there seems to be only the result of Redheffer available.

Similar inequalities have been studied for a more restricted class of functions y satisfying certain end-point conditions by Pfeffer [1974] and Fink [1977].

Section 2. The basic problem here is to characterize the function classes $W(p, K, J)$ and $W(p, J)$. In other words the problem is to find "useful," necessary and sufficient conditions on the weight function w for (3.53) to hold either for a given constant K or for some $K > 0$. This seems to be a rather difficult problem.

Sections 3 and 4. The crucial result in Section 3 is Theorem 3.7, due to Kwong and Zettl [1974a]. The remaining theorems are basically applications of Theorem 3.7 and previous results. An exception to this statement is Theorem 3.10 — also due to Kwong and Zettl [1981b]. Results in Section 4 are also from the same paper.

Section 5. The situation in which the weight function vanishes at some point is complicated. The results of Goldstein, Kwong, and Zettl [1983] surveyed in this section only touch on some of the simpler questions.

Chapter 4

The Difference Operator

In this chapter the difference operator is considered on the classical l^p spaces. Let $Z = \{\ldots - 2, -1, 0, 1, 2, \ldots\}$, and let $Z^+ = \{0, 1, 2, \ldots\}$. With $M = Z$ or $M = Z^+$ the difference operator Δ is defined by

$$x = (x_m)_{m \in M}, \quad \Delta x = (x_m - x_{m-1})_{m \in M},$$

$$\Delta^2 = \Delta(\Delta), \quad \Delta^j = \Delta(\Delta^{j-1}), \; j = 2, 3, \ldots$$

In contrast with the derivative operator $D = d/dt$, Δ is a bounded linear operator defined on all of l^p and is onto l^p for any p, $1 \le p \le \infty$, and $M = Z$ or $M = Z^+$.

4.1 The Discrete Product Inequality

The discrete analogue of the inequality (1.25) has apparently only quite recently been established.

Theorem 4.1 *(Kwong and Zettl [1988]). Let $M = Z$ or $M = Z^+$. Let k, n be integers with $1 \le k < n$. Let $1 \le p, r \le \infty$. There is a positive number C such that*

$$\|\Delta^k x\|_q \le C \, \|x\|_p^\alpha \, \|\Delta^n x\|_r^\beta \tag{4.1}$$

holds for all x in $l^p(M)$ satisfying $\Delta^n x \in l^r(M)$ if and only if

$$nq^{-1} \le (n-k)p^{-1} + kr^{-1} \tag{4.2}$$

and

$$\alpha = (n - k - r^{-1} + q^{-1})/(n - r^{-1} + p^{-1}), \beta = 1 - \alpha. \tag{4.3}$$

Remark 4.1 Given p, q, r, with $1 \leq p$, q, $r \leq \infty$, n, k, α, β satisfying the conditions of Theorem 4.1, there is clearly a smallest positive constant C such that (4.1) holds for all $x \in l^p(M)$ with $\Delta^n x \in l^r(M)$. This smallest (i.e., best) constant is denoted by $C = C(n, k, p, q, r, M)$ to emphasize its dependence on these quantities. We also let $C(n, k, p, M) = C(n, k, p, p, p, M)$. Surprisingly, there is less known about these discrete constants C than there is about their continuous analogues K.

Proof. The proof of the continuous analogue of (4.1), namely inequality (1.25), does not seem to extend to the discrete case. Here we use the continuous result to prove the discrete result.

Using standard approximation arguments it can be shown that

$$C(n, k, q, p, r, Z_{Z+}) \geq K(n, k, q, p, r, R_{R+}). \tag{4.4}$$

This is done in Ditzian [1983], see also Kaper-Spellman [1987], for the case $q = p = r = \infty$ with Z and R. Since the proof for the general case is similar, we omit the details. Thus, if either (4.2) or (4.3) fails, then $C \geq K = \infty$. By this we mean that (1.25), and consequently (4.1), is not valid. Assume conditions (4.2) and (4.3) are satisfied. We will show that (1.25) implies (4.1). For this it is sufficient to prove the case $n = 2$ since $n > 2$ then follows by induction. (The induction argument is not completely straightforward—see Kwong and Zettl [1980a] for details.) We proceed with the "whole line" version of $n = 2$, i.e., $M = Z$. The case $M = Z^+$ is similar and hence omitted.

To relate the discrete case to the continuous case we use a construction due to Ditzian [1983].

Given a sequence $x = \{x_j\}_{j \in M}$, define a function $f = Tx$ on R by

$$f(t) = \sum_{j \in M} x_j B_{j,3}(t), \quad t \in R \tag{4.5}$$

where $B_{j,3}$ is the B-spline of order 3 with support in $[j, j+3]$. See [de Boor [1978], Chapter IX] for a discussion of B-splines. Then f' is the piecewise linear interpolant of Δx and f'' is the piecewise constant interpolant of $\Delta^2 x$ with constant values of -1 and $+1$:

$$\begin{aligned}
f'(t) &= \sum_{j \in M} (\Delta x)_{j-1} B_{j,2}(t), \\
f''(t) &= \sum_{j \in M} (\Delta^2 x)_{j-2} B_{j,1}(t).
\end{aligned}$$

Thus, if $\Delta^2 x \in l^r(Z)$, then $f'' \in L^r(R)$ and

$$\|f''\|_r = \|\Delta^2 x\|_r. \tag{4.6}$$

Now we show that there is a positive number A, independent of x and f, such that

$$\|f\|_p \leq A \, \|x\|_p. \tag{4.7}$$

Note that each of $B_{j,3}(t)$ is a translate of $B_{0,3}(t)$ and has support in an interval of length 3. Let M be a bound of $B_{0,3}(t)$. Then for $1 \leq p < \infty$,

$$\|f\|_p^p = \| \sum_{j=-\infty}^{\infty} x_j B_{j,3} \|_p^p \leq \sum_{j=-\infty}^{\infty} \int_j^{j+1} |x_j B_{j,3}(t) + x_{j+1} B_{j+1,3}(t) + x_{j+2} B_{j+2,3}(t)|^p dt$$

$$\leq \sum_{j=-\infty}^{\infty} M^p |x_j + x_{j+1} + x_{j+2}|^p \leq 3 \cdot M^p 2^{p-2} \sum_{j=-\infty}^{\infty} |x_j|^p = 3 \cdot 2^{p-1} M^p \|x\|_p^p.$$

The proof of the $p = \infty$ case is similar.

Next we show that there is a positive number B, independent of x and f, such that

$$\|f'\|_q \geq B \, \|\Delta x\|_q. \tag{4.8}$$

Since f' is the piecewise linear function joining the points $(n, (\Delta x)_n)$, we need only show, when $1 \leq q < \infty$, that for some positive number D we have

$$\int_j^{j+1} |f'(t)|^q dt \geq D \left[|(\Delta x)_j|^q + |(\Delta x)_{j+1}|^q \right]. \tag{4.9}$$

Let $a = (\Delta x)_j, b = (\Delta x)_{j+1}$.

Case 1. The numbers a and b have the same sign. Suppose $a \geq 0$, $b \geq 0$. Let f_1 denote the straight line through the points $(j, \ a)$ and $(j + 1/2, \ 0)$ and let f_2 denote the straight line through the points $(j + 1/2, \ 0)$ and $(j + 1, \ b)$. Then

$$\int_j^{j+1} |f'|^q \geq \int_j^{j+1/2} |f_1|^q + \int_{j+1/2}^{j+1} |f_2|^q = \frac{a^q}{2(q+1)} + \frac{b^q}{2(q+1)}.$$

A similar construction establishes the case when $a \leq 0$ and $b \leq 0$. Hence (4.8) holds with $B^q = \dfrac{1}{2(q+1)}$ in this case.

Case 2. The numbers a and b have opposite signs. Suppose $a \geq 0$ and $b \leq 0$ and $a \geq |b|$. Let g denote the straight line through the points $(j, \ a)$ and $(j + 1/2, \ 0)$. Then

$$\int_j^{j+1} |f'|^q \geq \int_j^{j+1/2} |g|^q = \frac{a^q}{2(q+1)} \geq \frac{a^q + |b|^q}{4(q+1)}.$$

In the last step we used $|b| \leq a$. Clearly (4.8) follows from these inequalities.

The other subcases are established similarly, as is the case $q = \infty$.

Using (4.6), (4.7), and (4.8), we have that for all x in $l^p(Z)$ such that $x \neq 0$ and $0 \neq \Delta^2 x$ is in $l^r(Z)$,

$$\frac{\|\Delta x\|_q}{\|x\|_p^\alpha \|\Delta^2 x\|_r^\beta} \leq B^{-1} \cdot A \cdot \frac{\|f'\|_q}{\|f\|_p^\alpha \|f''\|_r^\beta} \leq AB^{-1}K. \tag{4.10}$$

This completes the proof of Theorem 4.1. $\square$

Remark 4.2 Inequality (4.10) yields an upper bound for C in terms of K:

$$C(2, 1, q, p, r, Z) \leq B^{-1}AK(2, 1, q, p, r, R). \tag{4.11}$$

Similar upper bounds for C in terms of K follow for $n > 2$ and all k, $1 \leq k < n$. However, these upper bounds for C are rough. We do not pursue the question of improving these bounds here.

4.2   The Second Order Case

In this section we investigate the special case $n = 2, k = 1, p = q = r$ of (4.1) more closely. Let $C(p, M) = C(2, 1, p, p, p, M)$, $K(p, J) = K(2, 1, p, p, p, J)$. Even in these special cases the best constants are known explicitly only when $p = 1, 2, \infty$ in both cases $M = Z$ or $M = Z^+$ and $J = R$ or $J = R^+$. These results are summarized in

Theorem 4.2 $C(1, Z) = \sqrt{2} = K(1, R)$
$$C(2, Z) = 1 = K(2, R)$$
$$C(\infty, Z) = \sqrt{2} = K(\infty, R)$$
$$C(1, Z^+) = \sqrt{5/2} = K(1, R^+)$$
$$C(2, Z^+) = \sqrt{2} = K(2, R^+)$$
$$C(\infty, Z^+) = 2 = K(\infty, R^+)$$

Proof. Proofs of these results can be found in the references given in Kwong and Zettl [1980b]. $\square$

It is interesting to note that the discrete constants are the same as the corresponding continuous constants in all six cases where they are known explicitly. Thus, one might be tempted to conjecture that $C(p, Z) = K(p, R)$ and $C(p, Z^+) = K(p, R^+)$ for all p, $1 \leq p \leq \infty$. That this is not so, at least for the whole line case, is shown by

Theorem 4.3 *(Kwong and Zettl [1988]). For some values of $p > 3$ we have $C(p, Z) > K(p, R)$.*

In proving Theorem 4.3, we first establish

Lemma 4.1 *For $1 \leq p \leq 2$ we have*

$$C(p,\ Z) \geq 2^{2/p-1} \tag{4.12}$$

and for $2 \leq p < \infty$ we have

$$C(p,\ Z) \geq 2^{1-2/p}. \tag{4.13}$$

Proof. (of Lemma 4.1) A sequence $x = (x_j)$ is said to be P-periodic if P is a positive integer such that for all j in Z

$$x_{j+P} = x_j.$$

For such a sequence x we define its "periodic l^p norm" as

$$\|x\|_{p,P} = \left(\sum_{j=0}^{P-1} |x_j|^p \right)^{1/p},\ 1 \leq p < \infty.$$

Note that if x is P-periodic then so is Δx.

By Theorem 8 of Kwong and Zettl [1987] we have

$$C(p,Z) = \sup \frac{\|\Delta x\|_{p,P}}{\|x\|_{p,P}\,\|\Delta^2 x\|_{p,P}} \tag{4.14}$$

where the supremum is taken over all nonzero P-periodic sequences x in $l^\infty(Z)$ with $\Delta^2 x \neq 0$ for all $P = 1, 2, 3, \ldots$

Applying (4.14) to the 4-periodic sequences $\ldots 0\ 1\ 0\ -1\ \ldots$ and $\ldots 1\ 1\ -1\ -1\ldots$, we get (4.12) and (4.13), respectively. (Actually, both (4.12) and (4.13) hold for all p, $1 \leq p < \infty$ but are interesting only for the ranges of p indicated.) $\square$

It was shown in Theorem 2.11 of Chapter 2 that

$$K(p,R) \leq U(p)$$

where

$$U(p) = (q-1)^{(2-q^n)q^{-n}} \left(\Pi_{i=1}^n \left(\frac{q}{q^i-1} - 1 \right)^{q-i} \right)^{2(q-1)}, \tag{4.15}$$

$2 < p < \infty, p^{-1} + q^{-1} = 1, n = [(\log_2 q)^{-1}]$ and $[\cdot]$ is the greatest integer function.

From (4.13) and (4.15), (see also [Franco, Kaper, Kwong and Zettl [1983], p. 261]) for $p = 4$, we get

$$K(4,\ R) \leq U(4) = (15/7)^{3/8} \doteq 1.33082962 < 2^{1/2} \doteq 1.414214 \leq C(4,Z).$$

Similarly, we obtain

$$K(5,R) \leq U(5) \;=\; 4^{47/125}(11/9)^{8/25}(19/61)^{32/125}$$

$$\doteq 1.33222966 \;<\; 2^{1-2/5} \doteq 1.515717 \leq C(5,Z).$$

$$K(6,R) \leq U(6) \;=\; 5^{19/108}(19/11)^{5/18}(59/91)^{25/108}$$

$$\doteq 1.39745611 \;<\; 2^{1-2/6} \doteq 1.587401 \leq C(6,Z).$$

Remark 4.3 Numerical evidence strongly suggests that $C(p,\,Z) > K(p,\,R)$ for all p satisfying $3 < p < \infty$. In fact this follows from the lower bound (4.13) for $C(p,\,Z)$ and the upper bound $U(p)$ for $K(p,\,R)$ for every p in the range $3 < p < \infty$ for which we have made the computation including values of p up to $p = 10^5$.

Question 1.   Is $C(p,\,Z) > K(p,\,R)$ for $1 < p < 2$ and for $2 < p < \infty$?

It seems to us that the upper bound $U(p)$ of $K(p,\,R)$ is "good" when $p > 3$ but not so good when $p < 3$. We expect that before Question 1 is answered for $1 < p < 2$ and $2 < p < 3$, a better upper bound than $U(p)$ needs to be found.

The "half line" version of Question 1 is:

Question 2.   Is $C(p,\,Z^+) > K(p,\,R^+)$ for $1 < p < 2$ and $2 < p < \infty$?

In attempting to answer Question 2, one is hampered by the lack of "good" upper bounds for $K(p,\,R^+)$ and "good" lower bounds for $C(p,\,Z^+)$.

There seem to be no upper bounds comparable to $U(p)$ known for $K(p,\,R^+)$ and no lower bounds comparable to (4.12), (4.13) known for $C(p,\,Z^+)$.

Can the lower bounds (4.12), (4.13) be improved? or

Question 3.   Is $C(p,\,Z) = 2^{1-2p}$ for $2 < p < \infty$?

Question 4.   Is $C(p,\,Z) = 2^{2/p-1}$ for $1 < p < 2$?

4.3   Extremals

Here we study the question of extremals for the inequality

$$\|\Delta x\|_p^2 \le C(p,\ M)\,\|x\|_p\,\|\Delta^2 x\|_p. \tag{4.16}$$

An extremal for inequality (4.16) is a sequence x in $l^p(M)$ which is not identically zero for which equality holds in (4.16). Such an x is also called an extremal for the constant $C(p,\ M)$. In [Copson 1979] it was shown, with an "elementary" proof, that there are no extremals for $C(2,\ Z^+)$ and $C(2,\ Z)$. On the other hand, it is easy to check that

$$x = (\ldots 1,\ 1,\ -1,\ -1,\ 1,\ 1,\ -1,\ -1,\ldots)$$

is an extremal of $C(\infty, Z)$. There are many other extremals for this case, e.g.,

$$x = (\ldots 1,\ 1,\ 1,\ -1,\ -1,\ -1,\ 1,\ 1,\ 1,\ -1,\ -1,\ -1,\ldots).$$

Below we show that there are no extremals for $C(1,\ Z^+), C(1,\ Z)$, and $C(\infty,\ Z^+)$.

Clearly, the constant C can be characterized as follows:

$$C(p,\ M) = \sup \frac{\|\Delta x\|_p^2}{\|x\|_p\|\Delta^2 x\|_p},$$

where the supremum is taken over all nonzero sequences x in $l^p(M)$. Let

$$Q(x) = Q(p,\ M,\ x) = \frac{\|\Delta x\|_p^2}{\|x\|_p\,\|\Delta^2 x\|_p}, \qquad x \in l^p(M), x \ne 0.$$

Below, in a proof given when p and M are fixed, we use $Q(x)$ in place of $Q(p,\ M,\ x)$ and $\|\ \|$ instead of $\|\ \|_p$.

Theorem 4.4 *There is no extremal for* $C(1,\ Z^+)$.

Proof. Suppose $x = (x_j)_{j=0}^\infty$ is such an extremal.

Case 1. Assume $x_0 = 0$.   Let

$$y = (\ldots - x_2, -x_1,\ 0,\ x_1,\ x_2,\ \ldots).$$

Then y is in $l^1(Z)$ and a straightforward computation, using the fact that $C(1,\ Z) = 2$, yields

$$\frac{5}{2} = \frac{\|\Delta x\|^2}{\|x\|\|\Delta^2 x\|} = \frac{\|\Delta y\|^2}{\|y\|\|\Delta^2 y\|} \le 2.$$

This contradiction shows that $x_0 = 0$ is impossible.

Case 2. Assume $x_0 \neq 0$.   Consider

$$y = (x_0, \ (x_0 + x_1)/2, \ x_1, \ (x_1 + x_2)/2, \ x_2, \ldots)$$

Then y is in $l^1(N)$ and

$$
\begin{aligned}
\Delta y &= (x_1 - x_0, \ x_1 - x_0, \ x_2 - x_1, \ x_2 - x_1, \ x_3 - x_2, \ldots) \\
\Delta^2 y &= (0, \ x_2 - 2x_1 + x_0, \ 0, \ x_3 - 2x_2 + x_1, \ldots)/2
\end{aligned}
$$

Thus,

$$\|\Delta y\| \ = \ \|\Delta x\|, \ \|\Delta^2 y\| \ = \ \|\Delta^2 x\|/2$$

and

$$
\begin{aligned}
\|y\| &= \ \|x\| + (|x_0 + x_1| + |x_1 + x_2| + \ldots)/2 \\
&\leq \ \|x\| + (\|x\| + |x_1| + |x_2| + \ldots)/2 < 2\,\|x\|,
\end{aligned}
$$

since $x_j = 0$ for all $j \geq 1$ is impossible, because this would give $Q(x) = 1$. Consequently, $Q(y) > Q(x)$, contradicting the assumption that x is an extremal. This completes the proof of Theorem 4.4. $\square$

Theorem 4.5 *There is no extremal for* $C(\infty, \ Z^+)$.

The proof of Theorem 4.5 depends on several lemmas and the introduction of the concept of "turning point".

Definition. A sequence

$$x = (x_0, \ x_1, \ x_2, \ldots)$$

is said to have a *turning point* at x_i, $i = 1, \ 2, \ldots$ if either

$$x_i \geq x_{i-1} \text{ and } x_i \geq x_{i+1}$$

or

$$x_i \leq x_{i-1} \text{ and } x_i \leq x_{i+1}.$$

Lemma 4.2 *If* x *in* $l^\infty(Z^+)$ *has a turning point at* x_i, *then*

$$\left(\sup_{j \geq i} |(\Delta x)_j|\right)^2 \leq 2\,\|x\|_\infty\,\|\Delta^2 x\|_\infty.$$

Proof. Let

$$y = (\ldots x_{i+2},\ x_{i+1},\ x_i,\ x_i,\ x_{i+1}, x_{i+2}, \ldots)$$

Then y is in $l^\infty(Z)$ and

$$\|y\| \le \|x\|, \ \|\Delta y\| = \sup_{j \ge i} |(\Delta x)_j|, \ \|\Delta^2 y\| \le 2 \ \|x\| \ \|\Delta^2 x\|.$$

The result now follows from Theorem 4.2 of section 2. □

Note the following consequence of Lemma 4.2.

Corollary 4.1 *If x in $l^\infty(Z^+)$ has the property that*

$$\|\Delta x\|^2 > 2\|x\| \ \|\Delta^2 x\|,$$

then either x has no turning point or

$$\|\Delta x\| = |x_j| \ \text{for some} \ j < i$$

where x_i is the first turning point. In particular this is true for an extremal x of $C(\infty, Z^+)$.

Lemma 4.3 *If x is an extremal of either $C(\infty, Z^+)$ or $C(\infty, Z)$, then*

$$\sup(x_j) = \|x\| \ \text{and} \ \inf(x_j) = -\|x\|.$$

Proof. Suppose

$$\epsilon = \|x\| - \sup(x_j) > 0.$$

Then the sequence $y_i = x_j + \epsilon/2$ has the property that

$$\|y\| = \|x\| - \epsilon/2, \ \|\Delta y\| = \|\Delta x\| \ \text{and} \ \|\Delta^2 y\| = \|\Delta^2 x\|.$$

Consequently, $Q(y) > Q(x)$, contradicting the fact that x is an extremal. Replace x by $-x$ to obtain the other half of the lemma. □

Lemma 4.4 *Every extremal of $C(\infty, Z^+)$ has a turning point.*

Proof. Suppose x is an extremal for $C(\infty, N)$ with no turning point. Then x must be strictly monotone. By replacing x with $-x$, if necessary, we can assume, without loss of generality, that x is strictly decreasing. By Lemma 4.3

$$x_0 = \|x\|_\infty = -\lim x_j = -c > -\infty, \ \text{as} \ j \to \infty.$$

It follows that $(\Delta x)_j \to 0$ and $(\Delta^2 x)_j \to 0$ as $j \to \infty$. Thus, there exists an integer J such that for all $j > J$ we have $|(\Delta x)_j| < \|\Delta x\|_\infty$ and $|(\Delta^2 x)_j| < \|\Delta^2 x\|_\infty$. Therefore, $\|\Delta x\|_\infty = |(\Delta x)_j|$ for some $j \leq J$. Let

$$y = (x_0, \ x_1, \ x_2, \dots \ x_J, \ x_J, \ x_J, \dots).$$

Then y is in $l^\infty(N)$ and $\|y\| = \|x\|$, $\|\Delta y\| = \|\Delta x\|$, $\|\Delta^2 y\| \leq \|\Delta^2 x\|$. Therefore $Q(y) \geq Q(x)$ and y is also an extremal. But this contradicts Lemma 4.3 since

$$\inf_i(y_i) = y_J = x_J < -\|x\|_\infty. \ \Box$$

Lemma 4.5 *If x is an extremal for $C(\infty, Z^+)$ and x_i is its first turning point, then the sequence*

$$y = (x_0, \ x_1, \dots, \ x_{i-1}, \ x_i, \ x_i, \ x_i, \dots)$$

is also an extremal.

Proof. By Lemma 4.1, $\|\Delta x\| = |x_j|$ for some $j < i$. Thus $\|\Delta y\| = \|\Delta x\|$. Clearly $\|y\| \leq \|x\|$ and $\|\Delta^2 y\| \leq \|\Delta^2 x\|$. Hence $Q(y) \geq Q(x)$ and y must be an extremal. By replacing y with $-y$, if necessary, we may assume that

$$x_0 > x_1 > x_2 > \dots > x_{i-1} > x_i = x_{i+1} = x_{i+2} = \dots \qquad . \Box$$

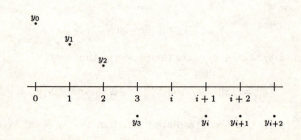

Lemma 4.6 *Suppose x is an extremal of $C(\infty, Z^+)$. Let y be the extremal constructed in Lemma 4.5. Then*

$$\|\Delta y\| = y_0 - y_1 > y_j - y_{j+1}, \ j = 1, 2, 3, \dots \qquad (4.17)$$

Proof. Since Δy has only a finite number of nonzero terms, we have

$$\|\Delta y\| = |y_{j+1} - y_j| \text{ for some } j. \qquad (4.18)$$

Let j denote the largest integer for which (4.18) holds. Define

$$z = (y_j, \ y_{j+1}, \ y_{j+2}, \ldots) = (z_0, \ z_1, \ z_2, \ldots)$$

Then

$$\|\Delta y\| = \|\Delta z\|, \ \|\Delta^2 z\| \leq \|\Delta^2 y\|, \ \|z\| \leq \|y\|$$

unless $j = 0$. Thus, $j > 0$ implies $Q(z) > Q(y)$. But this is impossible, since y is an extremal. Therefore, $j = 0$ and the proof of Lemma 4.6 is complete. $\square$

Lemma 4.7 *If $C(\infty, Z^+)$ has an extremal, then it has an extremal $y = (y_0, \ y_1, \ y_2, \ldots)$ satisfying*

$$(\Delta^2 y)_j \geq 0, \ j = 0, \ 1, \ 2, \ldots \tag{4.19}$$

Proof. We start with a definition.

Definition. The sequence $x = (x_0, \ x_1, \ x_2, \ldots)$ is said to be

(i) concave up at x_i, $i > 0$ if

$$(\Delta^2 x)_i > 0.$$

(ii) concave down at x_i, $i > 0$ if

$$(\Delta^2 x)_i < 0,$$

(iii) straight at x_i, $i > 0$ if

$$(\Delta^2 x)_i = 0.$$

An extremal y of the form $y = (y_0, \ y_1, \ldots, \ y_n, \ y_n, \ y_n, \ldots)$ is concave up at y_1, since

$$\|\Delta y\| = y_0 - y_1 > y_j - y_{j+1}, \ j = 1, \ 2, \ 3, \ldots$$

and concave up at y_n, because by Lemma 4.4, $y_0 > y_1 > y_2 > \ldots y_n = y_{n+1} = \ldots$.

For $j = 0$, (4.19) follows from Lemma 4.6.

If $(\Delta^2 y)_1 < 0$, construct a new sequence z by moving y_2 down onto the line going through y_1 and y_3. If $(\Delta^2 z)_2 \geq 0$, we do not change y_3 and we proceed as above, but if $(\Delta^2 z)_2 < 0$, we return y_2 to its original position, construct the line through y_1 and y_4, then move both y_2 and y_3 onto this line. If $(\Delta^2 z)_3 \geq 0$, we proceed as before. However, if $(\Delta^2 z)_3 < 0$, we change the previous construction by moving y_2 and y_3 both to their original positions, and then moving y_2, y_3 and y_4 all down onto the line through y_1 and y_5. Continuing this process we have, after a finite number of steps, constructed a sequence z satisfying (4.19). Note that

$$\|z\| = \|y\|, \|\Delta z\| = \|\Delta y\|, \text{ and } \|\Delta^2 z\| \leq \|\Delta^2 y\|.$$

Thus, $Q(z) \geq Q(y)$, and therefore z must also be an extremal. This completes the proof of Lemma 4.7. $\square$

Lemma 4.8 *Suppose*

$$y = (y_0, \ y_1, \ y_2, \ldots, \ y_n, \ y_n, \ y_n, \ldots) \tag{4.20}$$

is an extremal satisfying (4.17), (4.18), (4.19) with

$$y_0 > y_1 > y_2 > \ldots > y_n = y_{n+1} = \ldots$$

Then

$$(\Delta^2 y)_j = \|\Delta^2 y\| = m, \ j = 0, \ 1, \ 2, \ \ldots, \ n-1. \tag{4.21}$$

Proof. If $(\Delta^2 y)_{j-1} < m$ for some $j = 1, \ 2, \ \ldots, \ n$, then construct a new sequence z by replacing y_j by $y_j - \epsilon$. For sufficiently small positive ϵ, we then have

$$(\Delta^2 z)_{j-1} < m, \ \ (\Delta^2 z)_i \leq (\Delta^2 y)_i$$

for $i = j - 1$ and $i = j$, and

$$\|z\| = \|y\|, \ \ \|\Delta z\| \leq \|\Delta y\|.$$

Repeating this, if necessary, results in a new sequence z satisfying $Q(z) > Q(y)$, contradicting the extremality of y. $\square$

Proof. (of Theorem 4.5) Suppose $C(\infty, Z^+)$ has an extremal. Then there is an extremal y of the form (4.20) satisfying (4.17), (4.18), (4.19), (4.21). Then

$$r = y_{n-1} - y_n = (\Delta^2 y)_{n-1} \leq m.$$

By induction using (4.21), we establish that

$$y_{n-2} - y_n = 2r + m,$$
$$y_{n-3} - y_n = 3r + 3m,$$
$$y_0 - y_n = nr + n(n-1)m/2.$$

By Lemmas 4.2 and 4.5 we have

$$\|y\| = (y_0 - y_n)/2; \ \ \|\Delta y\| = y_0 - y_1 = r + (n-1)m.$$

Let $\alpha = r/m \leq 1$. Then

$$Q(y) \ = \ \frac{\|\Delta y\|^2}{\|y\| \ \|\Delta^2 y\|} \ = \ \frac{2(r + (n-1)m)^2}{m(nr + n(n-1)m/2)}$$

$$= \ \frac{2(\alpha + (n-1))^2}{n(\alpha) + n(n-1)/2} \ = \ \frac{4(\alpha + (n-1))^2}{2n(\alpha) + n(n-1)} < 4,$$

since

$$(\alpha + n - 1)^2 = \alpha^2 + 2(\alpha)(n-1) + (n-1)^2 = \alpha^2 + 2n(\alpha) - 2(\alpha) + n(n-1) - n - 1$$
$$= 2n(\alpha) + n(n-1) - (n - 1 + 2(\alpha) - (\alpha)^2) < 2n(\alpha) + n(n-1).$$

This contradiction completes the proof of Theorem 4.5. □

Theorem 4.6 *There is no extremal for* $C(1, Z)$.

The proof of Theorem 4.6 is based on several lemmas. First we introduce some notation and definitions.

Given a sequence $x = (x_j)_{-\infty}^{\infty}$, construct a new sequence $y = Ax$ as follows:

$$y = (\ldots x_{-2}, (x_{-2} + x_{-1})/2, x_{-1}, (x_{-1} + x_0)/2, x_0, (x_0 + x_1)/2, x_1, (x_1 + x_2)/2, x_2, \ldots).$$

Then y is in $l^1(Z)$ if x is in $l^1(Z)$ and

$$\|y\| = \|x\| + (\ldots + |x_{-2} + x_{-1}| + |x_{-1} + x_0| + |x_0 + x_1| + |x_1 + x_2| + \ldots)/2$$
$$\leq \|x\| + (\|x\| + \|x\|)/2 = 2\|x\|,$$

with equality holding if and only if x_j and x_{j+1} have the same sign for all $j = 0, -1, +1, -2, +2, \ldots$

$$\Delta y = (\ldots, (x_{-1} - x_{-2})/2, (x_{-1} + x_{-2})/2, (x_0 - x_{-1})/2, (x_0 - x_{-1})/2, (x_1 - x_0)/2, \ldots)$$
$$\Delta^2 y = (\ldots, 0, (x_0 - 2x_{-1} + x_{-2})/2, 0, (x_1 - 2x_0 + x_{-1})/2, 0, \ldots)$$
$$\|\Delta y\| = \|\Delta x\|, \quad \|\Delta^2 y\| = \|\Delta^2 x\|/2.$$

Hence

$$Q(y) \geq Q(x) \tag{4.22}$$

and equality holds if and only if

$$x_j x_{j+1} \geq 0 \text{ for all } j = 0, -1, 1, -2, 2, \ldots \tag{4.23}$$

Lemma 4.9 *If x is an extremal of $C(1, Z)$, then $y = Ax$ is also an extremal and any two adjacent terms of x must have the same sign. In particular, x can change sign only by going through a zero.*

Proof. This follows from (4.22) and (4.23). □

Note that between any two consecutive zeros of x, all terms (if there are any) must have the same sign.

Below, restrictions of a sequence x to finite or infinite "intervals" of Z play an important role. Such restrictions of x will be called "subsections" of x.

Definition. A subsection of x in $l^p(Z)$ is a restriction of x to a finite or infinite interval of Z.

It is convenient to introduce some notation. For any n in Z, x_n^L and x_n^R denote the restrictions of x to $(-\infty, n]$ and $[n, \infty)$, respectively. Specifically,

$$x_n^L = (\ldots, \; x_{n-2}, \; x_{n-1}, \; x_n), \; x_n^R = (x_n, x_{n+1}, \; x_{n+2}, \ldots).$$

Note that x_n is common to both subsections.

Given n, m in Z with $-\infty < m < n < \infty$, the restriction of x to the interval $[m, n]$ is denoted by $x_{m,n}$. Thus

$$x_{m, \; n} = (x_m, \; x_{m+1}, \ldots, x_{n-1}, x_n).$$

Next we define "norms" of subsections of a bi-infinite sequence x. These are not norms in the usual linear space sense. Nevertheless, it is convenient to use the norm terminology and notation for our purposes here. To avoid confusion with the usual norms, we use the symbol $||| \; |||$.

$$||| x_n^L ||| \; = \; |x_n|/2 + \sum_{j=\infty}^{n-1} |x_j|, \;\; ||| x_n^R ||| = |x_n|/2 + \sum_{j=n+1}^{\infty} |x_j|,$$

$$||| x_{m,n} ||| \; = \; (|x_m| + |x_n|)/2 + \sum_{j=m-1}^{n-1} |x_j|.$$

Note that

$$||x|| = ||| x_n^L ||| + ||| x_n^R |||, \;\; ||x|| = ||| x_m^L ||| + ||| x_{m,n} ||| + ||| x_n^R |||.$$

Given a subsection s of x we wish to consider Δs and $\Delta^2 s$. These are defined in the usual way: Δs is the collection of forward differences of s; $\Delta^2 s$ is the set of forward differences of Δs. If s is a finite subsection of x with k terms then Δs has $k - 1$ terms and $\Delta^2 s$ has $k - 2$ terms. When $s = (x_m, \; x_{m+1}), \Delta s = (x_{m+1} - x_m)$ and $\Delta^2 s$ is vacuous. Note that Δs is a subsection of Δx but s and Δs are defined on different subintervals of Z.

The "norm" of Δs is defined as follows: Let s be any subsection of x. Let

$$||| \Delta s ||| = \sum |x_{j+1} - x_j|$$

where the summation extends over all j such that both x_j and x_{j+1} are terms of s.

The "norm" of $\Delta^2 s$ is defined in a more complicated way, depending on the nature of the finite endpoints of s. Let

$$|||\Delta^2 s||| = \sum |(\Delta^2 s)_j| + u$$

where the summation extends over all indices j such that $(\Delta^2 s)_j$ is defined, i.e., such that all three terms x_{j+2}, x_{j+1}, x_j are members of s; and

1. $u = 0$ if all finite endpoints of s are zero, i.e., $s = s_n^R$ or s_n^L and $x_b = 0$ or $s = x_{m,n}$ and $x_m = 0 = x_n$;

2. if $s = s_n^L$ and x_n is not $= 0$ then $u = |x_{n+1} - x_n|$,

   if $s = s_n^R$ and x_n is not $= 0$ then $u = |x_n - x_{n-1}|$,

   if $s = s_{m,n}$ and both x_m, x_n are not 0 then $u = |x_{m+1} - x_m| + |x_n - x_{n-1}|$,

   if $s = s_{m,n}$ and $x_n = 0$ but $x_m \neq 0$, then $u = |x_{m+1} - x_m|$,

   if $s = s_{m,n}$ and $x_m = 0$ but x_n is not 0 then $u = |x_n - x_{n-1}|$.

For any subsection s of x we define

$$Q(s) = \frac{|||\Delta s|||^2}{|||s||| \; |||\Delta^2 s|||}.$$

Observe that for any x in $l^1(Z)$ with terms x_m, x_n, $m < n$ which are either zero or turning terms we have

$$\begin{aligned} ||\Delta x|| &= |||x_n^L||| + |||x_n^R|||, \\ ||\Delta x|| &= |||x_m^L||| + |||\Delta x_{m,n}||| + |||\Delta x_m^R|||, \end{aligned}$$

and if $x_n = 0$, then

$$||\Delta^2 x|| = |||\Delta^2 x_n^L||| + |||\Delta^2 x_n^R||| + |x_{n+1} + x_{n-1}|.$$

If x_n is not 0 but is a turning term, then

$$\begin{aligned} ||\Delta^2 x|| &= |||\Delta^2 x_n^L||| + |||\Delta^2 x_n^R|||, \\ ||\Delta^2 x|| &= |||\Delta^2 x_m^L||| + |||\Delta^2 x_{m,n}||| + |||\Delta^2 x_n^R||| + u, \end{aligned}$$

$$\text{where } u = \begin{cases} |x_{m+1} + x_{m-1}| \text{ if } x_m = 0; \\ |x_{n+1} + x_{n-1}| \text{ if } x_n = 0; \\ \text{the sum of these two quantities if both } x_n, x_m \text{ are } 0; \\ 0 \text{ if both of } x_m, x_n \text{ are nonzero turning points;} \\ |x_{m+1} + x_{m-1}| \text{ if } x_m \text{ is a turning point and } x_n = 0; \\ |x_{n+1} + x_{n-1}| \text{ if } x_n = 0 \text{ and } x_m \text{ is a turning point.} \end{cases}$$

In any case we have

$$\|\Delta^2 x\| \geq \||\Delta^2 x_n^L\|| + \||\Delta^2 x_n^R\||$$
$$\|\Delta^2 x\| \geq \||\Delta^2 x_m^L\|| + \||\Delta^2 x_{m,n}\|| + \||\Delta^2 x_n^R\||.$$

Below we will need the additional observation that for any subsections of x in $l^1(Z)$, say s_1 and s_2, which have only an endpoint in common, say x_n, if s is the subsection whose terms consist of all the terms of s_1 and s_2, then

$$\||s\|| = \||s_1\|| + \||s_2\||, \quad \||\Delta s\|| = \||\Delta s_1\|| + \||\Delta s_2\||,$$
$$\||\Delta^2 s\|| = \||\Delta^2 s_1\|| + \||\Delta^2 s_2\|| + u,$$

where $u = 0$ if x_n is a nonzero turning point, and $u = |x_{n+1} + x_{n-1}|$ if $x_n = 0$.

Again, in either case, we have

$$\||\Delta^2 s\|| \geq \||\Delta^2 s_1\|| + \||\Delta^2 s_2\||.$$

Lemma 4.10 *Assume x is an extremal of $C(1, Z)$. If s is any subsection of x such that any endpoint of s is either zero or a nonzero turning point of x, then*

$$Q(s) \leq Q(x) = C(1, Z).$$

Proof. The proof is divided into several cases depending on the nature of the subsection s.

Case 1. $s = x_n^L$ or $s = x_n^R$ and $x_n = 0$. Without loss of generality we may assume that $n = 0$. Suppose $s = x_0^R$. Let

$$y = (\ldots, -x_2, -x_1, 0, x_1, x_2, \ldots).$$

Then y is in $l^1(Z)$ and a straight forward computation shows that

$$Q(s) = Q(y) \leq C(1, Z) = Q(x).$$

For $s = x_0^L$ let

$$y = (\ldots, x_{-2}, x_{-1}, 0, -x_{-1}, -x_{-2}, \ldots)$$

and proceed as above.

Case 2. $s = x_n^L$ or $s = x_n^R$ and x_n is a nonzero turning point of x. We may assume that $n = 0$. If $s = x_0^R$, let

$$y = (\ldots, x_{-2}, x_{-1}, x_0, x_{-1}, x_{-2}, \ldots).$$

Then y is in $l^1(Z)$ and a simple computation, using the definition of $|\!|\!|s|\!|\!|$ given above, shows that

$$\|\Delta^i y\| = 2|\!|\!|\Delta^i s|\!|\!|, i = 0, 1, 2.$$

Hence

$$Q(s) = Q(y) \leq C(1, Z) = Q(x).$$

Case 3. $s = x_{m,n}$ with $-\infty < m < n < \infty$.

(a) $x_m = 0 = x_n$. Let y be the periodic sequence whose period y_P is given by:

$$y_P = (0, -x_{n-1}, \ldots, -x_{m+2}, -x_{m+1}, 0, x_{m+1}, x_{m+2}, \ldots,$$
$$0, -x_{n-1}, -x_{n-2}, \ldots, -x_{m+1}, 0).$$

Then with $r = 3(n - m) + 1$ we get

$$Q(s) = \frac{\|\Delta y\|_{1,r}^2}{\|y\|_{1,r} \, \|\Delta^2 y\|_{1,r}} \leq C(1, Z) = Q(x)$$

where the inequality follows from [Kwong and Zettl 1989].

(b) $x_m = 0$ and x_n is a nonzero turning term. Here we construct a periodic sequence whose period consists of the finite sequence:

$$(0 = x_m, x_{m+1}, x_{m+2}, \ldots, x_n, x_{n-1}, x_{n-2}, \ldots, x_{m+1}, x_m = 0)$$

and proceed as in Case 3 (a).

(c) x_m is a nonzero turning term and x_n is zero. This case is similar to 3(b).

(d) Each of x_m and x_n is a nonzero turning point. We construct an infinite periodic sequence y with period y_P given by:

$$y_P = (x_n, x_{n-1}, \ldots, x_{m+1}, x_m, x_{m+1}, \ldots, x_{n-1}, x_n, x_{n-1}, \ldots, x_{m+1}, x_m)$$

and proceed as in Case 3(a). This completes the proof of Lemma 4.10. $\square$

Lemma 4.11 *Suppose s_1 and s_2 are subsections of an extremal x of $C(1, Z)$ which have only an endpoint in common. Then*

$$Q(s_1) = Q(s_2) = Q(x) = 2 = C(1, \ Z)$$

and the two three-vectors

$$(|||s_1|||, |||\Delta s_1|||, |||\Delta^2 s_1|||), (|||s_2|||, |||\Delta s_2|||, |||\Delta^2 s_2|||)$$

are linearly dependent.

Proof. First we establish the special case when $s_1 = x_n^L$ and $s_2 = x_n^R$. Let

$$A = |||s_1|||, \; B = |||\Delta^2 s_1|||, \; C = |||s_2|||, \; D = |||\Delta^2 s_2|||.$$

Then

$$\|x\| = A + C, \; \|\Delta x\| = |||\Delta s_1||| + |||\Delta s_2|||, \; \|\Delta s_2\| \geq B + D.$$

¿From Lemma 4.10 and the Schwarz inequality in R_2 we obtain

$$\|\Delta x\|^2 = (|||\Delta s_1||| + |||\Delta s_2|||)^2 \leq 2((AB)^{\frac{1}{2}} + (CD)^{\frac{1}{2}})^2$$
$$\leq 2(A + C)(B + D) \leq 2\|x\|\|\Delta^2 x\| = \|\Delta x\|^2.$$

Hence we must have equality throughout. Thus,

$$Q(s_1) = Q(x) = Q(s_2); x_{n+1} + x_{n-1} = 0 \text{ if } x_n = 0;$$
$$A = kC \text{ and } B = kD \text{ for some constant } k.$$

¿From this it follows that $\|\Delta s_1\| = k\|\Delta s_2\|$ and the proof is complete for this special case.

Now suppose $s_1 = x_{m,n}$. From the special case established above we know that $Q(s_2) = Q(x)$ for $s_2 = x_m^R$. Now let $s_2 = x_n^R$ and proceed as in the special case above with x replaced by x_m^R. (Note that $x_m^R = x_{m,n} + x_n^R$ where by '+' here we mean the subsection obtained by combining all the terms of $x_{m,n}$ and x_n^R.) We may then conclude that

$$Q(x_{m,n}) = Q(x_n^R) = Q(x).$$

The other cases are established similarly. The linear dependence of the two three-vectors in the general case also follows from the special case established above. This completes the proof of Lemma 4.10. $\square$

Corollary 4.2 *If the extremal x has a zero at x_n then*

$$x_{n+1} + x_{n-1} = 0. \tag{4.24}$$

In particular this means that x cannot have two consecutive zeros. Another consequence of (4.24) is the fact that no turning term can be a zero.

Lemma 4.12 *Suppose $x = (x_j)$ is an extremal of $C(1, Z)$ and x_m, x_n with $m < n$ are two turning points of x. Then $x_j = 0$ for some j such that $m < j < n$.*

Proof. Assume that x has no zero between x_m and x_n. By replacing x by $-x$, if necessary we may assume that $x_j > 0$ for $m < j < n$. From Corollary 4.2 we may conclude that $x_m > 0$ and $x_n > 0$. Let $s = x_{m, n}$. By Lemma 4.11, $Q(s) = Q(x) = C(1, Z)$. For $\epsilon > 0$ consider the finite sequence

$$s(\epsilon) = (x_m - \epsilon, x_{m+1} - \epsilon, \ldots, x_n - \epsilon).$$

Then proceeding as in the proof of Lemma 4.10, case 3(d), one can show that $Q(s(\epsilon)) \leq Q(x) = C(1, Z)$. (Note that the assumption that x is an extremal is not needed in this proof.) On the other hand for ϵ sufficiently small we obtain $Q(s(\epsilon)) > Q(s) = Q(x)$. This contradiction establishes Lemma 4.11. $\square$

Corollary 4.3 *Suppose x is an extremal of $C(1, Z)$. Then there is exactly one turning point between any two consecutive zeros of x.*

Proof. This follows immediately from Lemma 4.12. $\square$

Lemma 4.13 *Assume x is an extremal of $C(1, Z)$ having consecutive zeros at x_m, x_n and a turning point at x_r with $m < r < n$. Replacing x by $-x$, if necessary, we can assume that $x_r > 0$. Thus by Corollary 4.3 $x = (x_j)$ is increasing for $m \leq j \leq r$ and decreasing for $r \leq j \leq n$. Then all x_j for $m \leq j \leq r$ lie on the same straight line.*

Proof. Let s denote the subsection

$$s = (x_m, x_{m+1}, \ldots, x_r).$$

Suppose that for some j satisfying $m < j < r$ we have

$$x_{j-1} - 2x_j + x_{j+1} < 0.$$

Let

$$t = (x_m, \ldots, x_{j-1}, (x_{j-1} + x_j)/2, x_{j+1}, \ldots, x_r).$$

Then

$$|||t||| \; < \; |||s|||, \quad |||\Delta t||| = |||\Delta s|||, \quad |||\Delta^2 t||| \; < \; |||\Delta^2 s|||.$$

Hence

$$Q(t) > Q(s) = Q(x) = C(1, Z).$$

This leads to a contradiction, since it can be shown, using the proof of Lemma 4.9, Case 3(b), that $Q(t) \leq Q(x)$. Note that, although the definition of the $\|\| \ \|\|$ is given only for subsections of x it can be applied to nonsubsections of x and the proof of Lemma 4.9 also applies to subsections of elements x of $l^\infty(Z)$ which are not extremals.

Thus the graph of s must be convex. If it does not lie on a straight line then there is a smallest integer j between m and r such that

$$x_{j-1} - 2x_j + x_{j+1} > 0.$$

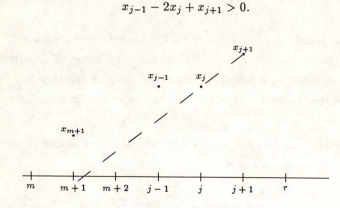

Construct the straight line L through the two points (j, x_j) and $(j+1, x_{j+1})$.

Case 1. The Line L intersects the x-axis to the right of $m + 1$. Let

$$t = (0 = t_0, t_1, \ldots, t_{i-1} = x_j, t_i = x_{j+1})$$

where the $t's$ are all chosen to lie on the straight line through x_j and x_{j+1}. Then

$$\|\|t\|\| \ < \ \|\|s\|\|, \ \|\|\Delta t\|\| \ = \ \|\|\Delta s\|\|, \ \|\|\Delta^2 t\|\| \ < \ \|\|\Delta^2 s\|\|$$

and therefore $Q(t) > Q(s)$ leading to a contradiction as before.

Case 2. Suppose L intersects the x-axis at u with $m < u \leq m + 1$. We reduce Case 2 to Case 1 as follows. The sequence Ax constructed above (immediately after the statement of Theorem 4.6) is also extremal. Replacing x by Ax halves the distance between the x-coordinates of consecutive terms of x. Repeating this, if necessary, we conclude that for some positive integer q, $A^q x$ is in Case 1 above.

Similarly, we can show that the terms of the subsection

$$u = (x_r, \ x_{r+1}, \ldots, \ x_n)$$

lie on the same decreasing straight line.

By Lemma 4.10, $Q(s) = Q(u)$, and the two triples

$$(|||s|||, |||\Delta s|||, |||\Delta^2 s|||), \quad (|||u|||, |||\Delta u|||, |||\Delta^2 u|||),$$

are proportional. But $|||\Delta s||| = |||\Delta u|||$ and therefore $|||s||| = |||u|||$ and $|||\Delta^2 s||| = |||\Delta^2 u|||$. This means that the lines connecting the points of the subsection s form an isosceles triangle.

Next we show that x cannot have a largest zero. Suppose $x_m = 0$ and $x_j > 0$ for $m < j$. (If all $x_j > 0$ for $j > m$ replace x by $-x$). There must be a turning term to the right of x_m, say, x_t. Proceeding as above we can show that the subsection from x_m to x_t lies on an increasing straight line and the subsection from x_t onward to the right lies on a decreasing straight line. But this is impossible since x is in $l^1(Z)$. □

Proof. (of Theorem 4.6) To complete the proof of Theorem 4.6 recall that it follows from what we have shown above that any extremal of $C(1, Z)$ must consist of subsections whose points lie on an infinite number of isosceles triangles. In addition, by Corollary 4.2 we have that $x_{j+1} + x_{j-1} = 0$ whenever $x_j = 0$. Thus all these triangles have the same base angle. This is impossible since x is in $l^1(Z)$. This completes the proof of Theorem 4.6. □

4.4 Notes and Problems

Section 1. The special case $p = 1 = r$ of Theorem 4.1 is due to Ditzian [1983]. The proof for the general case given here is taken from Kwong and Zettl [1988].

Besides those already raised in the text, there are many interesting open questions about the constants C, particularly regarding their relationship with their continuous analogues K. We mention a few below.

An obvious question is

Question 1. Does equality hold in (4.4)? In other words is it true that, given p, q, r, n, k satisfying (4.2) with α, β given by (4.3), we have

$$C(n, k, p, q, r, Z) = K(n, k, p, q, r, R)?$$

and

$$C(n, k, p, q, r, Z^+) = K(n, k, p, q, r, R^+)?$$

Ditzian [1983] showed that the answer, in general, is no. He found that

$$C(3, 1, \infty, Z) = 2^{\frac{1}{3}} > (9/8)^{\frac{1}{3}} = K(3, 1, \infty, R). \tag{4.25}$$

Below we summarize the known cases of equality and inequality in (4.4) as far as we are aware of them. Proofs can be found in the references given for each case.

I. $$C(n, k, \infty, Z^+) = K(n, k, \infty, R^+), \quad 1 \leq k < n. \tag{4.26}$$

Equality (4.26) follows from an abstract result. If A is an m-dissipative linear, but not necessarily bounded, operator on a Banach space X, then

$$\|A^k x\| \leq K(n, k, \infty, R^+) \|x\|^{(n-k)/n} \|A^n x\|^{k/n}, \quad x \in D(A^n), \ 1 \leq k < n. \tag{4.27}$$

A linear (bounded or unbounded) operator A on a Banach space X is m-dissipative if and only if it is the infinitesimal generator of a strongly continuous C_0 contraction semigroup. If A is the infinitesimal generator of a strongly continuous <u>group</u> of isometries then (4.27) can be strengthened to

$$\|A^k x\| \leq K(n, k, \infty, R) \|x\|^{(n-k)/n} \|A^n x\|^{k/n}, \quad x \in D(A^n), \quad 1 \leq k < n. \tag{4.28}$$

Inequalities (4.27) and (4.28) are established in Certain and Kurtz [1975]. The case $n = 2$ was previously established by Kallman and Rota [1971].

If X is a Hilbert space then (4.27) can be improved to

$$\|A^k x\| \leq K(n, k, 2, R^+) \|x\|^{(n-k)/n} \|A^n x\|^{k/n}, \ x \in D(A^n), \ 1 \leq k < n. \tag{4.29}$$

The case $n = 2$ of (4.29) was proven by Kato [1971] and the general case was proven independently by Chernoff [1979], Phong [1981] and Kwong and Zettl [1979].

Since Δ is an m-dissipative operator on any of the spaces $l^p(M), 1 \leq p \leq \infty, M = Z$ or $M = Z^+$, the inequalities (4.4), (4.27), (4.28), (4.29) yield some information about the constants $C(n, k, p, M)$. In particular equality (4.26) follows from (4.4) and (4.27).

II. $$C(n, k, 2, Z^+) = K(n, k, 2, R^+), \quad 1 \leq k < n. \tag{4.30}$$

Proof. This follows from (4.4) and (4.29). □

III. $$C(n, k, 2, Z) = K(n, k, 2, R^+) = 1, \ 1 \leq k < n. \tag{4.31}$$

Proof. The case $n = 2$ of the equality on the right is established in the classic book by Hardy, Littlewood and Polya [1934]. The general case follows by induction. The case $n = 2$ of the left equality was proven by Copson [1977] and the general case follows by induction. □

IV. $$C(n, \, n-1, \, \infty, Z) = K(n, \, n-1, \, \infty, R), \, n = 2, \, 3, \, 4, \ldots. \tag{4.32}$$

Proof. This was established by Ditzian [1983]. The proof depends on properties of B-splines. □

V. $$C(n, \ n-1, \ 1, Z) = K(n, \ n-1, \ 1, R), \ n = 2, \ 3, \ 4, \dots \qquad (4.33)$$

Proof. This follows from (4.4), (4.32), and (4.34). □

VI. $$K(n, k, 1, R) = K(n, k, \infty, R), \ 1 \le k < n. \qquad (4.34)$$

Proof. See Ditzian [1975]. □

VII. $$K(n, k, p, R^+) \le K(n, k, \infty, R^+), \ 1 \le p \le \infty. \qquad (4.35)$$

Proof. This is a consequence of the abstract inequality (4.27). □

VIII. $$K(n, k, p, R) \le K(n, k, \infty, R), 1 = p \le \infty. \qquad (4.36)$$

Proof. This is a consequence of (4.28). □

IX. $$C(n, k, p, Z^+) \le C(n, k, \infty, Z^+), \ 1 \le p \le \infty. \qquad (4.37)$$

Proof. This is a consequence of (4.27) and (4.26). See also Ditzian [1975] and Kwong and Zettl [1987]. □

X. $$C(n, k, p, Z) \le C(n, k, \infty, Z), \ 1 \le p \le \infty. \qquad (4.38)$$

Proof. In contrast with (4.36) this result does not seem to follow from the theory of semigroups of operators. A proof is given in Ditzian [1975]—see also Kwong and Zettl [1987]. □

The discrete analogue of (4.34) does not seem to be known. We raise:

Question 2. Is $C(n, k, 1, Z) = C(n, k, \infty, Z)$ for all n, k satisfying $1 \le k < n$? The answer is yes when $k = n - 1$. This follows from (4.33), (4.34) and (4.32). In addition the answer is also yes in all cases when both of these constants are known explicitly. See below.

Below we summarize cases for which the exact values of K or C are known explicitly. We do this only for the special case when $p = q = r$. See the survey paper by Kwong and Zettl [1980] for explicit values of K in some cases when not all three norms are the same. Explicit values of C do not seem to be known when not all three norms are equal. Here we use the notation $C(n, k, p, M) = C(n, k, p, p, p, M)$.

Known Cases

n	k	p	J or M	K or C	Author reference
2	1	∞	R^+	2	Landau [1913]
2	1	∞	Z^+	2	Gindler and Goldstein [1981]
2	1	∞	R	$2^{\frac{1}{2}}$	Hadamard [1914]
2	1	∞	Z	$2^{\frac{1}{2}}$	Ditzian and Neumann [1986]
					See also Kaper and Spellman [1987]
2	1	2	R^+	$2^{\frac{1}{2}}$	Hardy and Littlewood [1932]
2	1	2	Z^+	$2^{\frac{1}{2}}$	Copson [1979]
2	1	2	R	1	Hardy, Littlewood and Polya [1934]
2	1	2	Z	1	Copson [1979]
2	1	1	R^+	$(5/2)^{\frac{1}{2}}$	Berdyshev [1971]
2	1	1	Z^+	$(5/2)^{\frac{1}{2}}$	Kwong and Zettl [1986]
2	1	1	R	$2^{\frac{1}{2}}$	Ditzian [1975], Berdyshev [1971]
2	1	1	Z	$2^{\frac{1}{2}}$	Kwong and Zettl [1986]
3	1	∞	R	$(9/8)^{\frac{1}{3}}$	Shilov [1937]
*3	1	∞	Z	$2^{\frac{2}{3}}$	Ditzian and Neumann [1986]
3	2	∞	R	$3^{\frac{1}{3}}$	Shilov [1937]
3	2	∞	Z	$3^{\frac{1}{3}}$	Ditzian [1983]
4	1	∞	R	$(512/375)^{\frac{1}{4}}$	Shilov [1937]
*4	1	∞	Z	$2^{\frac{1}{2}}$	Ditzian and Neumann [1986]
					See also Kaper and Spellman [1987]
4	2	∞	R	$(36/25)^{\frac{1}{4}}$	Shilov [1937]
*4	2	∞	Z	$(4/3)^{\frac{1}{2}}$	Ditzian and Neumann [1986]
					See also Kaper and Spellman [1987]
4	3	∞	R	$(24/5)^{\frac{1}{4}}$	Shilov [1937]
4	3	∞	Z	$(24/5)^{\frac{1}{4}}$	Ditzian [1983]
5	1	∞	R	$(1953125/1572864)^{\frac{1}{5}}$	Kolmogorov [1938]
*5	1	∞	Z	$4^{\frac{1}{5}}$	Ditzian and Neumann [1986]
					See also Kaper and Spellman [1987]
5	2	∞	R	$(125/72)^{\frac{1}{5}}$	Shilov [1937]
*5	2	∞	Z	$(4/3)^{\frac{4}{5}}$	Ditzian and Neumann [1986]
					See also Kaper and Spellman [1987]

Known Cases

n	k	p	J or M	K or C	Author reference
5	3	∞	R	$(225/128)^{\frac{1}{5}}$	Kolmogorov
*5	3	∞	Z	$2^{\frac{1}{5}}$	Kwong and Zettl [1988]
5	4	∞	R	$(15/2)^{\frac{1}{5}}$	Kolmogorov [1938]
5	4	∞	Z	$(15/2)^{\frac{1}{5}}$	Ditzian [1983]
n	k	∞	R	K_1	Kolmogorov [1938]
n	n - 1	∞	Z	K_1	Ditzian [1983]
3	1	∞	R^+	$(243/8)^{\frac{1}{3}}$	Schoenberg and Cavaretta [1970]
3	2	∞	R^+	$(24)^{\frac{1}{3}}$	Schoenberg and Cavaretta [1970]
n	k	∞	R^+	K_2	Schoenberg and Cavaretta [1970]
n	k	∞	Z^+	K_2	Certain and Kurtz [1977]
					See also Gindler and Goldstein [1981]
n	k	2	R	1	Hardy, Littlewood and Polya [1934]
n	k	2	Z	1	Follows from special case $n=2$
					established by Copson [1979]
3	1	2	R^+	$3^{\frac{1}{2}}[2(2^{\frac{1}{2}}-1)]^{\frac{1}{3}}$	Ljubic [1960]
3	2	2	R	$3^{\frac{1}{2}}[2(2^{\frac{1}{2}}-l)]^{\frac{1}{3}}$	Ljubic [1960]
n	k	2	R^+	K_3	Ljubic [1960]
n	k	2	Z^+	$K(n,k,2,R^+)$	Chernoff [1979], Phong [1981],
					Kwong and Zettl [1979]
n	k	1	R	K_1	Ditzian [1975]
3	1	1	Z	$2^{\frac{1}{3}}$	Kwong and Zettl [1988]
3	2	1	Z	$3^{\frac{1}{3}}$	Kwong and Zettl [1988]
4	1	1	Z	$2^{\frac{1}{2}}$	Kwong and Zettl [1988]
4	2	1	Z	$(4/3)^{\frac{1}{2}}$	Kwong and Zettl [1988]
4	3	1	Z	$(24/5)^{\frac{1}{4}}$	Kwong and Zettl [1988]
5	1	1	Z	$4^{\frac{1}{5}}$	Kwong and Zettl [1988]
5	2	1	Z	$(4/3)^{\frac{4}{5}}$	Kwong and Zettl [1988]
5	3	1	Z	$2^{\frac{1}{5}}$	Kwong and Zettl [1988]
5	4	1	Z	$(15/2)^{\frac{1}{5}}$	Kwong and Zettl [1988]

Here

$$K_1 = K(n,k,\infty,R) = k_{n-k}k_n^{-(n-k)/n},$$

$$k_i = 4\pi^{-1}\sum_{j=0}^{\infty}(-1)^j(2j+1)^{-i-1} \text{ for } even \text{ } i$$

$$k_i = 4\pi^{-1}\sum_{j=0}^{\infty}(2j+1)^{-i-1} \text{ for } odd \text{ } i.$$

$K_2 = K(n,k,\infty,R^+)$: These constants have been characterized by Schoenberg and Cavaretta [1970] in terms of norms of Euler splines but an explicit formula for all n and k is not available.

$K_3 = K(n,k,2,R^+)$: An algorithm to compute these constants for all $k = 1, \ldots, n-1$, $n = 2, 3, \ldots$ was developed by Ljubic [1960]. This was refined elegantly by Kupcov [1977]. See Franco, Kaper, Kwong and Zettl [1985] for a numerical computation of K_3 and for some asymptotic estimates.

Comments on the above list of best constants

See Kwong and Zettl [1980b] for some information on best constants when not all three norms are equal. There are many gaps in our knowledge of K and C. We mention a few below.

The values of $K(n,k,1,R^+)$ seem to be known explicitly only for $n = 2$, $k = 1$. Note that although $K(n,k,1,R) = K(n,k,\infty,R)$ the corresponding result for R^+ is false: $(5/2)^{1/2} = K(2,1,1,R^+) \neq K(2,1,\infty,R^+) = 2^{1/2}$. Kwong and Zettl [1980a] proved that $K(2,1,p,R^+)$ is a continuous function of p. Hence $K(2,1,p,R^+) \neq K(2,1,q,R^+)$ where $p^{-1} + q^{-1} = 1$ for some values of p other than $p = 1$ and $p = \infty$.

The stars $*$ indicate cases when $C > K$. It is interesting to observe that, although $C(n, n-1, \infty, Z) = K(n, n-1, \infty, R)$ (and $C(n,k,\infty,Z^+) = K(n,k,\infty,R^+)$ for all n, k, $1 \leq k < n$), we have $C(n,k,\infty,Z) > K(n,k,\infty,R)$ in all cases when $1 \leq k < n-1$ for which these two constants are explicitly known. So the question arises:

Question 3. Is $C(n,k,\infty,Z) > K(n,k,\infty R)$ for all n, k satisfying $1 \leq k < n-1$?

A closely related question is:

Question 4. Is $C(n,k,1,Z) > K(n,k,1,R)$ for all n, k satisfying $1 \leq k < n-1$?

For general n,k satisfying $1 \leq k < n$, the constants $K(n,k,\infty,R)$, $K(n,k,2,R)$, $K(n,k,1,R)$, and $C(n,k,2,Z)$ are known explicitly; constants $K(n,k,\infty,R^+)$, $C(n,k,\infty,Z^+)$,

$K(n, k, 2, R^+)$ and $C(n, k, 2, Z^+)$ can be computed numerically with known algorithms; but $C(n, k, \infty, Z)$, $C(n, k, 1, Z)$, $K(n, k, 1, R^+)$ and $C(n, k, 1, Z^+)$ do not seem to be known.

Section 3. This section is taken from Kwong and Zettl [1987]. Are there extremals for $C(p, Z)$, $1 < p < \infty$, $p \neq 2$? for $C(p, Z^+)$, $1 < p < \infty$, $p \neq 2$?

Are there any extremals in the higher order case $n > 2$ for $C(n, k, p, M)$, $1 \leq p \leq \infty$, $M = Z$ or $M = Z^+$? The answer is quite likely to be yes for $C(n, k, \infty, Z)$. Is it yes for any other case?

References

R.A. Adams [1975], "Sobolev Spaces," Academic Press, New York (1975).

V.V. Arestov [1972], "Exact inequalities between norms of functions and their derivatives," Acta Scientiarum Mathematicarum, 33, 243-267 (1972).

V.V. Arestov [1967], "The best approximation to differentiation operators," Mathematicheskie Zametkie, 1, part 2, 149-154 (1967).

V.V. Arestov [1972], "On the best approximation of the operators of differentiation and related questions," Approximation Theory, Proceedings of Conference in Poznan, Poland. Reidel Publishing Co., Boston (1972).

V.I. Berdyshev [1971], "The best approximation in $L(0, \infty)$ to the differentiation operator," Mathematicheskie Zametki, 5, 477-481 (1971).

B. Bollabas and J.R. Partington [1984], "Inequalities for quadratic polynomials in hermitian and dissipative operators," Advances in Mathematics (1984), 51, 271-280 (1984).

Yu. G. Bosse (G.E. Shilov) [1937], "O neravenstvakh mezhdu proizvodnymi," Mosk. Univ. Sbornik raport nauchnykh studencheskikh kruzhkov, 17-27 (1937).

J. Bradley and W.N. Everitt [1974], "On the inequality $\|f''\|^2 \leq K\|f\|\|f^{(4)}\|$," Quart. J. Math. 25, 241-252 (1974).

J. Bradley and W.N. Everitt [1973], "Inequalities associated with regular and singular problems in the calculus of variations," Trans. Amer. Math. Soc. 182, 303-321 (1973).

K.W. Brodlie and W.N. Everitt [1975], "On an inequality of Hardy and Littlewood," Proc. Roy. Soc. Edinburgh A72, 179-186 (1975).

A.S. Cavaretta [1974], "An elementary proof of Kolmogorov's theorem," Amer. Math. Monthly 81, 480-486 (1974).

A.S. Cavaretta [1976], "One-sided inequalities for the successive derivatives of a function," Bull. Amer. Math. Soc. 82, 303-305 (1976).

A.S. Cavaretta [1976], "A refinement of Kolmogorov's inequality, " MRC Technical Summary Report #1788.

H. Cartan [1939], "On inequalities between the maxima of the successive derivatives of a function," Comptes Rendus Sci. Acad. 208, 414-426 (1939).

M.W. Certain and T.G. Kurtz [1977], "Landau-Kolmogorov inequalities for semigroups and groups," Proc. Amer. Math. Socl 63, 226-230 (1977).

Ceisielski and J. Musielak [1972], "Approximation Theory," Proceedings of the Conference jointly organized by the Mathematical Institute of the Polish Academy of Sciences and the Institute of Mathematics of the Adam Mickiewicz University held in Proznan 22-26 August, 1972. D. Reidel Publishing Co., Boston (1972).

E.T. Copson [1977], "On two integral inequalities," Proc. Roy. Soc. Edinburgh, 77A, 325-328 (1977).

E.T. Copson [1977], "On two inequalities of Brodlie and Everitt," Proc. Roy. Soc. Edinburgh, 77A, 329-333 (1977).

E.T. Copson [1979], "Two series inequalities," Proc. Roy. Soc. Edinburgh 83A, 109-114 (1979).

C. de Boor [1978], "A practical guide to splines;, Springer-Verlag, Berlin and New York.

Z. Ditzian [1975], "Some remarks on inequalities of Landau and Kolmogorov," Equ. Math. 12, 145-151 (1975).

Z. Ditzian [1977], "Note on Hille's question," Aequationes Mathematicae 15, 143-144 (1977).

Z. Ditzian [1983], "Discrete and shift Kolmogorov type inequalities," Proc. Roy. Soc. Edinburgh 93A, 307-317 (1983).

Z. Ditzian and D.J. Newman [1986], "Discrete Kolmogorov-type inequalities," preprint.

W.D. Evans and A. Zettl [1978], "Norm inequalities involving derivatives," Proc. Roy. Soc. Edinburgh, 82A, 51-70 (1978).

W.N. Everitt [1972], "On an extension to an integro-differential inequality of Hardy, Littlewood and Polya," Proc. Roy. Soc. Edinburgh (A) 69, 295-333 (1972).

W.N. Everitt and M. Giertz [1974], "On the integro-differential inequality $\|f'\|_2^2 \leq K\|f\|_p\|f''\|_q$," J. Math. Anal. and Appl. 45, 639-653 (1974).

W.N. Everitt and M. Giertz [1972], "Some inequalities associated with certain ordinary differential operators," Math. Z. 126, 308-326 (1972).

W.N. Everitt and M. Giertz [1974], "Inequalities and separation for certain ordinary differential operators," P. London Math. Soc. 3, 28, 352-372 (1974).

W.N. Everitt and D.S. Jones, "On an integral inequality," Proc. Roy. Soc. London (A), 357 (1977), 271-288.

W.N. Everitt and A. Zettl [1978], "On a class of integral inequalities," J. London Math. Soc. (2), 17, 291-303 (1978).

A. Friedman [1969], "Partial Differential Equations," New York (1969).

A.M. Fink [1977], "Best possible approximation constants," Trans. Amer. Math. Soc. 226, 243-255 (1977).

Z.M. Franco, H.G. Kaper, M.K. Kwong and Z. Zettl [1983], "Bounds for the best constant in Landau's inequality on the line," Proc. Roy. Soc. Edinburgh 95A, 257-262 (1983).

Z.M. Franco, H.G. Kaper, M.K. Kwong and A. Zettl [1983], "Bounds for the best constant in Landau's inequality on the line," Proc. Roy. Soc. Edinburg 95A, 257-262 (1983).

Z.M. Franco, H.G. Kaper, M.K. Kwong and A. Zettl [1985], "Best constants in norm inequalities for derivatives on a half-line," Proc. Roy. Soc. Edinburgh 100A, 67-84 (1985).

V.N. Gabushin [1967], "In equalities for norms of a function and its derivatives in L_p metrics," Mat. Zam. 1, 291-298 (1967).

V.N. Gabushin [1968], "Exact constants in inequalities between norms of derivatives of functions," Mat. Zam. 4, 221-232 (1968).

V.N. Gabushin [1969], "The best approximation for differentiation operators on the half-line," Math. Zam. 6, 573-582 (1969).

A. Gindler and J.A. Goldstein [1975], "Dissipative operator versions of some classical inequalities," J.D'Analyse Math. 28, 213-238 (1975).

A. Gindler and J.A. Goldstein [1981], "Dissipative operators and series inequalities," Bull. Austr. Math. Soc. 23, 429-442 (1981).

J. Goldstein, "On improving the constants in the Kolmogorov inequalities," preprint.

A. Gorny [1939], "Contributions to the study of differentiable functions of a real variable," Acta Math. 71, 217-358 (1939).

J. Hadamard [1914], "Sur le module maximum d'une fonction et de ses derivees," C.R. des Seances de l'annee 1914, Soc. Math. de France, 66-72 (1914).

G.H. Hardy and J.E. Littlewood [1932], "Some integral inequalities connected with the calculus of variations," Quart. J. Math. Oxford Ser. 2, 3, 241-252 (1932).

G.H. Hardy, J.E. Littlewood and G. Polya [1934], "Inequalities," Cambridge (1934).

E. Hille [1972], "Generalizations of Landau's inequality to linear operators," Linear operators and approximation (edited by P.L. Butzer, et al.), Birkhauser Verlag, Basel and Stuttgart (1972).

E. Hille [1970], "Remark on the Landau-Kallman-Rota inequality," Aequationes Mat. 4, 239-240 (1970).

E. Hille [1972], "On the Landau-Kallman-Rota inequality," J. of Approx. Theory 6, 117-122 (1972).

E. Hille and R.S. Phillips [1957], "Functional Analysis and Semigroups," Amer. Math. Soc. Coll. Publ. 31, Rev. ed. Providence (1957).

R.R. Kallman and G.C. Rota [1970], "On the inequality $|f'\|^2 \leq 4\|f\|\|f''\|$," Inequalities II (). Shisha, ed.), Academic Press, New York, 187-192 (1970).

H.G. Kaper and B.E. Spellman [1987], "Best constants in norm inequalities for the difference operator," Trans. of the Amer. Math. Soc. 299, No. 1 (1987).

T. Kato [1971], "On an inequality of Hardy, Littlewood and Polya," Advances in Math. 7, 217-218 (1971).

A.N. Kolmogorov [1962], "On inequalities between the upper bounds of the successive derivatives of an arbitrary function on an infinite interval," Amer. Math. Soc. transl. (1) 2, 233-243 (1962).

S. Kurepa [1970], "Remarks on the Landau inequality," Aequationes Math. 4, 240-241 (1970).

M.K. Kwong and A. Zettl [1979], "Remarks on best constants for norm inequalities among power of an operator," J. Approx. Theory, 26, 249-258 (1979).

M.K. Kwong and Z. Zettl [1979], "Norm inequalities for dissipative operators on inner product spaces," Houston J. Math 5, 543-557 (1979).

M.K. Kwong and A. Zettl [1980], "Ramifications of Landau's inequality," Proc. Roy. Soc. Edinburgh 86A, 175-212 (1980).

M.K. Kwong and A. Zettl [1980], "Norm inequalities for derivatives," Lecture Notes in Mathematics, Spring Verlag 846, 227-243 (1980).

M.K. Kwong and A. Zettl [1980], "Norm inequalities of product form in weighted L^p spaces."

M.K. Kwong and A. Zettl [1986], "Landau's inequality for the differential and difference operators," General inequalities V., Birkhauser Verlag, Basel. (to appear). Proceedings of the fifth International Conference on General Inequalities.

M.K. Kwong and A. Zettl [1987], "Best constants for discrete Kolmogorov inequalities," Houston J. Math. (to appear).

M.K. Kwong and A. Zettl [1988], "Landau's inequality for the difference operator," Broc. Amer. Math. Soc. (to appear).

M.K. Kwong and A. Zettl [1989], "Best constants for discrete Kolmogorov inequalities," Houston J. Math. (1989), 99-119.

E. Landau [1913], "Einige Ungleichungen fur zweimal differenzierbare Funktionen," Proc. London Math. 13, 43-49 (1913).

Ju I. Ljubic (or Yu Lyubich) [1960 and 1964], "On inequalities between the powers of a linear operator," Transl. Amer. Math. Soc. Ser. (2) 40 (1964), 39-84; translated from Irv. Akad. Nauk SSSR Ser. Mat. 24 (1960), 825-864.

G.G. Magaril-Il'jaev and V.M. Tihomirov [1981], "On the Kolmogorov inequality for fractional derivatives on the half-line," Analysis Mathematica 7, 37-47 (1981).

A.P. Matorin [1955], "Inequalities between the maximum absolute values of a function and its derivatives on the half-line," Ukrain. Mat. Zh. 7, 262-266 (1955).

D.S. Mitrinovic [1970], "Analytic Inequalities," Springer-Verlag, Berlin (1970).

L. Nirenberg [1955], "Remarks on strongly elliptic partial differential equations," Comm. Pure Appl. Math. 8, 648-674 (1955).

B. Sz. Nagy [1941], "Uber Integralungleichungen zwischen einer Funktion und ihrer Ableitung," Acta. Sci. Math. 10, 64-74 (1941).

B. Neta, "On determination of best possible constants in integral inequalities involving derivatives," preprint.

J.R. Partington [1981], "Hadamard-Landau inequalities in uniformly convex spaces," Math. Proc. Camb. Phil. Soc., 90, 259-264 (1981).

J.R. Partington [1979], "Constants relating a Hermitian operator and its square," Math. Proc. Camb. Phil. Soc. 85, 325-333(1979).

R.M. Redheffer [1963], "Uber eine beste Ungleichung zwischen den Normen von f, f', f''," Math. Zeitschr. 80, 390-397 (1963).

I.J. Schonberg [1973], "The elementary cases of Landau's problem of inequalities between ," Amer. Math. Monthly, 80, 121-158 (1973).

I.J. Schonberg and A. Cavaretta [1970], "Solution of Landau's problem concerning higher derivatives on the half line," MRCT. S.R. 1050, Madison, Wisconsin (1970).

G.E. Shilov [1937], "O neravenstvakh mezhdu proizvodnymi," Sbornik Rabot Studencheskikh Nauchnykh Kruzhov Moskovskogo Gosudarstvennogo Universiteta, 17-27 (1937).

S.B. Stechkin [1965], "Inequalities between norms of derivatives of an arbitrary function," Acta. Sci. Math. 26, 225-230 (1965).

S.B. Stechkin [1967], "The inequalities between upper bounds for the derivatives of an arbitrary function on the half-line," Mat. Zametki I V. 6, 665-674 (1967).

E.M. Stein [1957], "Functions of exponential type," Annals of Math. (2), 65, 582-592 (1957).

W. Trebels and V.I. Westphal [1972], "A note on the Landau-Kallman-Rota-Hille inequality," Linear Operators and Approximation (edited by P.L. Butzer et al), Birkhauser Verlag, Basel and Stuttgart (1972).

L.V. Taikov [1968], "Inequalities of Kolmogorov type and the best formulae for numerical differentiation," R. Mat. Zam. 4, 233-238 (1968).

V.G. Solyer [1976], "On an inequality between the norms of a function and its derivative," Izvestia Vysshikh Uebebaykh Zavendemy Matematika 2 (165) (1976).

Subject Index

Vol. 1483: E. Reithmeier, Periodic Solutions of Nonlinear Dynamical Systems. VI, 171 pages. 1991.

Vol. 1484: H. Delfs, Homology of Locally Semialgebraic Spaces. IX, 136 pages. 1991.

Vol. 1485: J. Azéma, P. A. Meyer, M. Yor (Eds.), Séminaire de Probabilités XXV. VIII, 440 pages. 1991.

Vol. 1486: L. Arnold, H. Crauel, J.-P. Eckmann (Eds.), Lyapunov Exponents. Proceedings, 1990. VIII, 365 pages. 1991.

Vol. 1487: E. Freitag, Singular Modular Forms and Theta Relations. VI, 172 pages. 1991.

Vol. 1488: A. Carboni, M. C. Pedicchio, G. Rosolini (Eds.), Category Theory. Proceedings, 1990. VII, 494 pages. 1991.

Vol. 1489: A. Mielke, Hamiltonian and Lagrangian Flows on Center Manifolds. X, 140 pages. 1991.

Vol. 1490: K. Metsch, Linear Spaces with Few Lines. XIII, 196 pages. 1991.

Vol. 1491: E. Lluis-Puebla, J.-L. Loday, H. Gillet, C. Soulé, V. Snaith, Higher Algebraic K-Theory: an overview. IX, 164 pages. 1992.

Vol. 1492: K. R. Wicks, Fractals and Hyperspaces. VIII, 168 pages. 1991.

Vol. 1493: E. Benoît (Ed.), Dynamic Bifurcations. Proceedings, Luminy 1990. VII, 219 pages. 1991.

Vol. 1494: M.-T. Cheng, X.-W. Zhou, D.-G. Deng (Eds.), Harmonic Analysis. Proceedings, 1988. IX, 226 pages. 1991.

Vol. 1495: J. M. Bony, G. Grubb, L. Hörmander, H. Komatsu, J. Sjöstrand, Microlocal Analysis and Applications. Montecatini Terme, 1989. Editors: L. Cattabriga, L. Rodino. VII, 349 pages. 1991.

Vol. 1496: C. Foias, B. Francis, J. W. Helton, H. Kwakernaak, J. B. Pearson, H∞-Control Theory. Como, 1990. Editors: E. Mosca, L. Pandolfi. VII, 336 pages. 1991.

Vol. 1497: G. T. Herman, A. K. Louis, F. Natterer (Eds.), Mathematical Methods in Tomography. Proceedings 1990. X, 268 pages. 1991.

Vol. 1498: R. Lang, Spectral Theory of Random Schrödinger Operators. X, 125 pages. 1991.

Vol. 1499: K. Taira, Boundary Value Problems and Markov Processes. IX, 132 pages. 1991.

Vol. 1500: J.-P. Serre, Lie Algebras and Lie Groups. VII, 168 pages. 1992.

Vol. 1501: A. De Masi, E. Presutti, Mathematical Methods for Hydrodynamic Limits. IX, 196 pages. 1991.

Vol. 1502: C. Simpson, Asymptotic Behavior of Monodromy. V, 139 pages. 1991.

Vol. 1503: S. Shokranian, The Selberg-Arthur Trace Formula (Lectures by J. Arthur). VII, 97 pages. 1991.

Vol. 1504: J. Cheeger, M. Gromov, C. Okonek, P. Pansu, Geometric Topology: Recent Developments. Editors: P. de Bartolomeis, F. Tricerri. VII, 197 pages. 1991.

Vol. 1505: K. Kajitani, T. Nishitani, The Hyperbolic Cauchy Problem. VII, 168 pages. 1991.

Vol. 1506: A. Buium, Differential Algebraic Groups of Finite Dimension. XV, 145 pages. 1992.

Vol. 1507: K. Hulek, T. Peternell, M. Schneider, F.-O. Schreyer (Eds.), Complex Algebraic Varieties. Proceedings, 1990. VII, 179 pages. 1992.

Vol. 1508: M. Vuorinen (Ed.), Quasiconformal Space Mappings. A Collection of Surveys 1960-1990. IX, 148 pages. 1992.

Vol. 1509: J. Aguadé, M. Castellet, F. R. Cohen (Eds.), Algebraic Topology - Homotopy and Group Cohomology. Proceedings, 1990. X, 330 pages. 1992.

Vol. 1510: P. P. Kulish (Ed.), Quantum Groups. Proceedings, 1990. XII, 398 pages. 1992.

Vol. 1511: B. S. Yadav, D. Singh (Eds.), Functional Analysis and Operator Theory. Proceedings, 1990. VIII, 223 pages. 1992.

Vol. 1512: L. M. Adleman, M.-D. A. Huang, Primality Testing and Abelian Varieties Over Finite Fields. VII, 142 pages. 1992.

Vol. 1513: L. S. Block, W. A. Coppel, Dynamics in One Dimension. VIII, 249 pages. 1992.

Vol. 1514: U. Krengel, K. Richter, V. Warstat (Eds.), Ergodic Theory and Related Topics III, Proceedings, 1990. VIII, 236 pages. 1992.

Vol. 1515: E. Ballico, F. Catanese, C. Ciliberto (Eds.), Classification of Irregular Varieties. Proceedings, 1990. VII, 149 pages. 1992.

Vol. 1516: R. A. Lorentz, Multivariate Birkhoff Interpolation. IX, 192 pages. 1992.

Vol. 1517: K. Keimel, W. Roth, Ordered Cones and Approximation. VI, 134 pages. 1992.

Vol. 1518: H. Stichtenoth, M. A. Tsfasman (Eds.), Coding Theory and Algebraic Geometry. Proceedings, 1991. VIII, 223 pages. 1992.

Vol. 1519: M. W. Short, The Primitive Soluble Permutation Groups of Degree less than 256. IX, 145 pages. 1992.

Vol. 1520: Yu. G. Borisovich, Yu. E. Gliklikh (Eds.), Global Analysis – Studies and Applications V. VII, 284 pages. 1992.

Vol. 1521: S. Busenberg, B. Forte, H. K. Kuiken, Mathematical Modelling of Industrial Process. Bari, 1990. Editors: V. Capasso, A. Fasano. VII, 162 pages. 1992.

Vol. 1522: J.-M. Delort, F. B. I. Transformation. VII, 101 pages. 1992.

Vol. 1523: W. Xue, Rings with Morita Duality. X, 168 pages. 1992.

Vol. 1524: M. Coste, L. Mahé, M.-F. Roy (Eds.), Real Algebraic Geometry. Proceedings, 1991. VIII, 418 pages. 1992.

Vol. 1525: C. Casacuberta, M. Castellet (Eds.), Mathematical Research Today and Tomorrow. VII, 112 pages. 1992.

Vol. 1526: J. Azéma, P. A. Meyer, M. Yor (Eds.), Séminaire de Probabilités XXVI. X, 633 pages. 1992.

Vol. 1527: M. I. Freidlin, J.-F. Le Gall, Ecole d'Eté de Probabilités de Saint-Flour XX – 1990. Editor: P. L. Hennequin. VIII, 244 pages. 1992.

Vol. 1528: G. Isac, Complementarity Problems. VI, 297 pages. 1992.

Vol. 1529: J. van Neerven, The Adjoint of a Semigroup of Linear Operators. X, 195 pages. 1992.

Vol. 1530: J. G. Heywood, K. Masuda, R. Rautmann, S. A. Solonnikov (Eds.), The Navier-Stokes Equations II – Theory and Numerical Methods. IX, 322 pages. 1992.

Vol. 1531: M. Stoer, Design of Survivable Networks. IV, 206 pages. 1992.

Vol. 1532: J. F. Colombeau, Multiplication of Distributions. X, 184 pages. 1992.

Vol. 1533: P. Jipsen, H. Rose, Varieties of Lattices. X, 162 pages. 1992.

Vol. 1534: C. Greither, Cyclic Galois Extensions of Commutative Rings. X, 145 pages. 1992.

Vol. 1535: A. B. Evans, Orthomorphism Graphs of Groups. VIII, 114 pages. 1992.

Vol. 1536: M. K. Kwong, A. Zettl, Norm Inequalities for Derivatives and Differences. VII, 150 pages. 1992.